D1826179

Legal issues and the Internet

Guideline

NOTE: Any references to the law are current at the date of issue of this Guideline publication.

LEEDS LIBRARY & IN... ...ON SERVICES
DISC... ...DED

CCTA
Central Computer and Telecommunications Agency

LONDON: HMSO

LD 0999914 0

Acknowledgements   The assistance of Bird & Bird, and in particular
Mark O'Conor, in the preparation of this volume
under contract to CCTA is gratefully acknowledged.

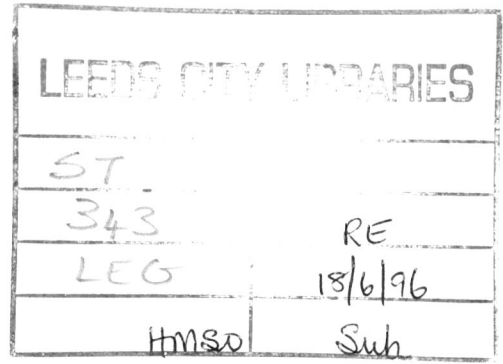

LEEDS CITY LIBRARIES

ST.
343   RE
LEG   18/6/96
        HMSO   Sub

© **Crown Copyright 1996**   Application for reproduction should be made to HMSO
Copyright Unit

ISBN 0 11 330682 2

For further information regarding this publication and other CCTA products please contact:

CCTA Library
Rosebery Court
St Andrews Business Park
Norwich NR7 0HS
01603 704930

Contents

Management summary

Today, people, companies and government organisations quite happily make use of the Internet to communicate, buy and sell and publish. The Internet is not "weird and wonderful"; it is a facility that, like many others in life, needs to be used sensibly and carefully within the applicable laws and regulations.

This *Guideline* is an introduction to using the Internet in a manner which is both prudent and legally sound. It examines some of the main ways in which business uses the Internet and the legal issues involved. As a companion to this *Guideline*, there is a *Reference Book* which provides an in-depth analysis of the principal laws and regulations which impact upon the use of the Internet with explanations using layman's terms as to how these laws and regulations are applied.

One of the popular uses of the Internet is to communicate by e-mail. Sending non-sensitive, informal information by e-mail has been likened to sending a postcard. However, just as we would send official mail more securely, so e-mails can be encrypted to provide confidentiality. Chapter 1, **Responsible use of e-mail**, provides guidance on the security and content of e-mails along with some practical advice on approaching the use of e-mail in a straightforward manner.

Individuals as well as organisations, in both the public and private sectors, can buy and sell products and services on the Internet somewhat akin to a mail order transaction. The World Wide Web (Web) provides an interactive means by which business can be transacted from the home or the office. Chapter 2, **Buying and selling on the Internet**, discusses contractual and payment arrangements, what evidence might be necessary and security aspects.

Information is "published" on the Internet through Web pages. It is common to draw the comparison between Web pages and noticeboards except that, in the case of the Internet, these noticeboards can be looked at from anywhere in the world. This global dimension means that Internet users need to give extra attention to the appropriateness of what information is "published". Chapter 3, **Publishing information**, provides advice and guidance on copyright in general and, for those users in the public sector, Crown and Parliamentary copyright, as well as the question of liability.

Chapter 4, **Undertaking procurement activities**, discusses the particular requirements for public sector procurement within the EU, where government organisations must ensure all those who tender for business are treated the same. It also addresses matters of wider interest to Internet users in general such as the widely used facilities of EDI.

Whether one is communicating, buying and selling, publishing or procuring on the Internet, there are some general actions and considerations that need to be taken by all users. In Chapter 5, **Connecting to and using the Internet**, and Chapter 6, **Setting up a Web page**, this *Guideline* provides some practical advice concerning these two subjects which can be addressed separately or in support of the information provided by the earlier chapters.

Finally, it is necessary to make it clear that all references to the law in both this *Guideline* and the *Reference Book* means the law of England and Wales. Although the majority of the references to the law in both publications applies to the United Kingdom as a whole, the law in Scotland and Northern Ireland does have subtle differences to the law of England and Wales. This means that anyone applying the information given in either publication within Scotland or Northern Ireland, will need to bear this in mind.

1   Responsible use of e-mail

1.1   Introduction

The Internet offers the potential for communication on a global basis. E-mail is a facility on the Internet that users of computer systems are likely to be familiar with already and is one aspect of the Internet which is likely to be used increasingly as a low cost way of improving communication between organisations.

Routing of e-mail is unpredictable, and a message may pass through many computers, in various countries, before arriving at its destination. In simple terms it works due to the nature of the "packet switched network". This means that a message (or series of packets) takes whatever route is available to its destination at any one time. For example, at 8 am a message from London to Edinburgh may be routed via the US, whereas at 8 pm it may be routed via Manchester.

E-mail is complementary to other forms of communication, but there are advantages over traditional methods, some of which are:

- an e-mail may be sent to as many recipients as required at the same time without any degradation in readability or completeness

- it is easier and more convenient to send attachments with e-mail such as files, software or even video clips

- attachments, for example a draft agreement, may go back and forth between two parties several times in a day, being amended at each end, which has obvious economic advantages when compared with the delays associated with traditional methods of delivery

- electronic debates or brainstorming sessions can be held swiftly and succinctly and comments can be passed around quickly to others often saving hours of time before a face-to-face meeting is needed.

It should also be remembered that there are risks associated with e-mail and these are touched upon in subsequent sections of this chapter.

1.2   Security

How to provide employees with the benefits of e-mail, without exposing organisations and their information systems to the many risks e-mail brings with it, is an important issue. Once an e-mail is sent, there is very little chance of retrieving it and it is almost effortless for the intended recipient to retransmit an e-mail message to another party. If a mistake is made the resultant damage may be magnified because of the possible widespread nature of the distribution. Organisations also need to be aware of the risks of deliberate or inadvertent disclosure of sensitive or confidential information to unauthorised parties by employees. Promoting security awareness and establishing proper procedures is essential (see 9.2.3 and 9.9 of *Reference Book*).

Unless it is encrypted, e-mail is less secure than a mobile phone or facsimile transmission. The message takes whatever route is available, which means that a sender has no control over the route, the number of countries (with differing legal systems) or the number of computers through which the message might pass. With a modest amount of IT expertise, it is relatively easy to intercept an e-mail message. It is possible to set up routines which scan passing e-mails for key words without being detected; the Internet equivalent of phone tapping. On the other side of the coin it is relatively easy to send an e-mail anonymously by directing it via an anonymous remailer which strips out all identifiers.

It is sensible to consider what types of communication are appropriate to send by e-mail and to consider carefully the content of e-mails.

1.2.1   Suitability of e-mail

E-mail may not be suitable for all types of communication. For example, it may not be prudent to send commercially confidential information by e-mail. For consistency, the definition of commercially confidential information for e-mail purposes should be tied to any existing classification of confidentiality within an organisation.

It may be advisable to specify that, for instance, no executable files should be sent as attachments and that documents containing start-up macros should not be transmitted, in order to avoid the inherent security risks.

1.2.2   Encryption

Encrypting an e-mail has been likened to sending a letter in a sealed envelope but this analogy is too simple. Sending *any* e-mail is comparable to using recorded delivery and encrypting the content is like writing in a language that only those who receive it understand. The use of encryption increases the potential for using Internet e-mail, especially in circumstances where the content is of a sensitive nature (see 9.9.3 of *Reference Book*).

1.3   **Content and types of communication**

The content of e-mails is subject to all applicable laws such as those relating to copyright, defamation, data protection and public records, as well as statutes concerning the sensitive and contentious issue of pornography. Obviously, nothing illegal or infringing a third party's intellectual property rights should be included in an e-mail.

Generally, the content determines whether e-mail is the most suitable mode of communication. However, there may well be a need to consider the legal aspects in detail or the particular implications of using the Internet. The *Reference Book* covers both the detail and the implications of the law for the Internet regarding copyright, defamation, data protection, public records and pornography in Chapters/Section 1/1.18, 2/2.5, 3/3.9, 4/4.3 and 5/5.8 respectively. It may well be necessary to also consider the eight principles of the Data Protection Act 1984 (see 3.3 of *Reference Book*) as well as the main regulatory environments (see Chapter 6 of *Reference Book*).

1.4   **Responsibility for sending e-mail**

If messages are transmitted on behalf of organisations, senders must make sure the content has been approved and that it is identifiable as either an official view or a private opinion.

Organisational or vicarious liability for e-mail sent by employees or, possibly, by others using the system with consent, could arise in a number of areas. Legally speaking, some of the areas of law which are most likely to impact upon e-mail use are copyright infringement, defamation, pornography, reliance on advisory content and contractual commitment. The nature and extent of the liability will vary in each case. In general terms, however, there are a number of common themes such as the vicarious liability of employers for the acts of their employees (see 9.7.1 of *Reference Book*).

Examination of the facts of each case will tend to show whether employees are liable on their own or whether the employer is liable under the principles of vicarious liability.

1.5	**Incoming e-mail**	Organisations should consider what steps to take, if any, with regard to incoming e-mail. Organisations may decide to monitor incoming e-mail in much the same way as telephone calls can be monitored. If e-mail is monitored, it is advisable to inform employees that monitoring is taking place. However, as employees do not have a legal right of privacy, strictly speaking, there is no need to do so.
1.6	**Evidential issues**	Authentication of senders and receivers is difficult and can raise evidential issues (see 7.5 of *Reference Book*). Good internal procedures should be maintained in order to prove, if need be, that a message was sent and that there are routines to authenticate the identities of senders and receivers. For example, where transmission or receipt may become an important issue, transmission and receipt reports should be retained in order to provide evidence that the message was sent.
1.7	**Copyright assignment**	As an item of property, a copyright work can be sold to another party, referred to as an "assignment of copyright", but statutory law states that an assignment must be in writing and "signed by or on behalf of the assignor" (see 1.11 of *Reference Book*). This in effect prevents a legally enforceable assignment occurring through an exchange of e-mails.
1.8	**Accessibility**	Employers should decide which employees can have access to the Internet, the extent of any supervision and any restrictions they wish to impose. For example, allowing only certain grades of staff to post a message or requiring individuals to seek approval from a higher

grade before posting a message; in much the same way as a draft letter might require approval from a superior before being sent out (see 9.2.3 of *Reference Book*).

Unnecessary, excessive or frivolous use of e-mail can be wasteful of time and money along with any associated drains on productivity.

1.9   Personal e-mail

Organisations should consider whether to allow employees to send personal e-mails although, in practice, it is probably difficult to control. If they are allowed, consider requiring employees to include at the foot of all personal e-mails, a specific disclaimer or an appropriate "signature" (a brief message appended automatically to the end of outgoing messages), to show that the message is not to be taken as being sent on behalf of the organisation. A simple disclaimer could follow the form of:

"This e-mail is a personal communication and is not authorised by or sent on behalf of any other person or company".

There is more advice on disclaimers at 6.2 of this *Guideline*.

1.10   Audit

Organisations should consider the need for putting an audit trail in place; for example, by copying certain categories of e-mail, such as those with contractual implications, onto paper and storing them on particular files, especially if required to do so under the Public Records Act 1958 (see Chapter 4).

1.11   Code of practice

It is recommended that organisations establish an e-mail code of practice covering the issues referred to in this *Guideline*. The code can be incorporated into an office manual or set of quality procedures. Alternatively, organisations may wish to consider incorporating it into an employee's contract of employment (see 9.8 of *Reference Book*).

2 Buying and selling on the Internet

2.1	**Introduction**	The introduction of the World Wide Web, often referred to as the Web, has meant that the commercial possibilities of the Internet have expanded greatly. Commercial organisations are waking up to the realisation that they have a whole new interactive way of reaching potential customers; customers who may be reached in the comfort of their own home without having to entice them into commercial premises through normal advertising and marketing practices.

One effective use of the Internet is for marketing and advertising but this is only half the story in the commercial sense. In order to harness fully the power of the Internet and its global interconnectivity, there is a requirement to be able to actually contract over the network. Traditional rules of contract must be extended for the purchase of products or the supply of services over the Internet.

Various estimates abound concerning the extent of investment in the Internet but it is generally accepted that between half and one billion dollars in cash or stock has changed hands over the last few years in buying and selling of Internet interests. More and more alliances are being formed between vendors, content owners and searchers etc. The rush of joint ventures and attempted joint ventures between telecommunications, computing and entertainment companies has coincided with the rise of the Internet as a commercial concern through the use of the Web.

Chapter 8 of the *Reference Book*, discusses the issues in greater depth.

2.2	**Contract terms**	It is by no means the norm for contracts to be in writing. Contracts arise daily and often the parties involved are unaware that a contract has been made at all.

There are generally two points at which written contract terms are examined. Firstly, in negotiation when determining exactly what the rights and obligations are going to be for each party. Secondly, when the relationship between the parties breaks down and it is necessary to determine who did or did not perform the contractual obligations.

There may be a problem in determining when a contract has been made, if at all, when the parties have communicated via the Internet. Unless it is made clear to the contrary, an offer on a Web page becomes a binding contract on receipt of a user response requesting to purchase products or services. This can create a problem for the selling organisation if an offer is left on a Web page beyond a date when it can be fulfilled.

To avoid this situation the selling organisation must make it clear that the offer on the Web page is merely an "invitation to treat". In this case the reader is invited to request to purchase (make the offer) and then the selling organisation can decide whether to accept (and make the contract) or reject the offer. Further advice on when a contract is made is given at 8.2.1 of *Reference Book*.

Due to the global nature of the Internet, difficulties can also arise in deciding where a contract is made, if at all; for example, whether an e-mail between the US and the UK in which contractual obligations are offered and accepted is made under US or UK law. A good policy, particularly in high value contracts, is to specify "up front" which laws of which country apply to that particular contract (see Chapter 8 of *Reference Book* in general and 8.2.1 in particular).

The EU is currently in the final stages in the adoption of a Directive (Common position (EC) No 19/95; OJ No C288/1), on the protection of consumers in respect of distance contracts, which explicitly extends to the use of e-mail. When implemented, it requires those who supply products and services to provide certain information to the consumer prior to the conclusion of the contract. This includes the price, arrangements for payment, delivery and performance and the existence of a "right of withdrawal". Although this Directive is not yet law, organisations should be aware of their potential future obligations.

2.3   Evidential issues

Both contract creation and dispute resolution require evidence of the intent of the parties; but of course contracts made over the Internet are less tangible than contracts made in writing. Indeed, electronic contracts may never even reach paper form as to do so would defeat the object of paperless commerce. The ease of manipulation of electronic data gives rise to real concerns over the authentication of terms, proof of offer and acceptance, the authorisation of signatures to a contract, and indeed the ability to determine whether a contract has even come into being.

Evidential issues arise concerning documents which never take on paper form. Authenticity of electronic records is clearly a difficulty and procedures must be put in place to help prove the authenticity of a record if it comes to a dispute. Security measures such as passwords and access control are useful, as are logs which record the activities of the machine upon which the records are kept. Such issues are further complicated by scanned documents and hand written digitised signatures which can be appended automatically to documents. Digital (as opposed to digitised) signatures use public-key cryptography (pkc) for electronic authentication and non-repudiation purposes.

The ease with which electronic information can be manipulated means that clear internal procedures and guidelines need to be in place in order to provide a requisite level of evidential proof should the validity of a digital signature to a message be called into doubt (see 7.4 of *Reference Book*).

2.4   Payment

Payment for goods and services contracted for on the Internet can be:

- by traditional off-line methods (for example, sending cheques by post)

- using established methods on the Internet

- using new on-line methods.

Established methods include use of e-mail response forms in reply to Web site advertisements where credit card details are sent over the Internet. The risk is that, unless credit card details are encrypted then they may be read and abused by the controllers of servers en route or by hackers who obtain unauthorised access to the credit card details (see 8.3.2 of *Reference Book*).

Electronic cash (see 8.3.4 of *Reference Book*) is an example of a new on-line method of payment. In this situation a customer converts real money into "cyber-bucks" or some equivalent. A number of banks and other organisations are currently examining such possibilities. The prognosis is that as more goods and services become available over the Internet, the less need there will be to convert the electronic cash back to real money after each transaction.

It is not the intention of this *Guideline* to discuss the subject of charging for access to information held by the Internet. However, 8.4 of the *Reference Book* provides further information on the subject.

2.5   Encryption

Encryption essentially scrambles plain text into an unintelligible code for transmission purposes allowing the receiver, by use of the requisite key, to unscramble the message. Encryption should play a key role in making business more secure over the Internet but encryption techniques vary and are only as good as their implementation.

There are various encryption-related services and technologies available for use on the Internet and these are referred to at 9.9.3 of the *Reference Book*.

The use of encryption should be tailored to the needs of individual organisations. Government use of cryptographic services to secure protected, marked information needs to be agreed with the security authorities.

2.6   Anonymity

It may be that, in certain circumstances, organisations wish to do business anonymously or under a pseudonym through the use of encryption such as public key directories or naming schemes. Studies are currently being carried out to look at the possibility of using encryption to enable an Internet user to do business under a pseudonym, which cannot be traced back to the true identity of the user. However, it is a common misconception that a user can move around the Internet without being seen; in fact the opposite is true. Whenever any Web pages are accessed, e-mails sent; or other communications or transactions made over the Internet, "electronic footprints" are very clearly left and this obviously raises serious issues in terms of privacy.

2.7   Trusted third parties (TTPs)

In a similar vein to the preceding section, it may be necessary in certain commercial circumstances to involve the use of TTPs in order to transact business over the Internet. TTPs include brokers or any similar intermediary through which goods or services may be procured. Care should be taken in selecting a reputable TTP. In addition it is important to have a clear understanding of both the limit of the authority of the TTP to act on the organisation's behalf, and the apportionment of liability, should the particular procurement of products and services in question become litigated. Good practice is to agree terms in advance.

2.8   Security for payment

There is little point in procuring products and services over the Internet if payment for those items cannot also be made over the Internet. Where payment details such as credit card numbers are to be sent over the Internet they should be encrypted. Note also that such information in conjunction with the name of the buyer constitutes "personal data" for the purposes of the Data Protection Act 1984. If this data is processed within the meaning of this Act, organisations need to comply with the requirements of the Act (see Chapter 3 of *Reference Book* for a general discussion of data protection issues and 3.3 in particular for the eight Principles of the Act).

An additional point to note is that new on-line methods of payment, such as electronic cash, are not currently backed by any bank and therefore are risky, especially for high value purchases.

| 2.9 | **Code of commerce** | There is currently no global, or even sectoral, code of commerce for doing business over the Internet. However, users, particularly those in various industry sectors, should look actively to the establishment of a code of commerce in order to conduct their business effectively and avoid the imposition of a code of commerce from above. For example codes exist which are used by various trading banks who trade using EDI and such codes may well prove a useful model for an Internet code of commerce (see 8.5.1, 10.3 and 10.4 of *Reference Book*). |

3   Publishing information

3.1   Introduction

Putting any information on the Internet may constitute publication to the world at large, and this raises certain issues which are considered in this chapter.

Using the Internet to publish information is having a profound effect upon the publishing industry. Magazines, journals and books made freely available over the Internet may have an impact on hard copy sales. However, hard copy may still be preferable for convenience and, if good quality copy is required, more cost-effective than downloading a copy from the Internet for printing.

The Internet opens up new possibilities for publishing information that either might not have been economically viable to publish in hard copy or, in the case of individuals, that a publisher might not have been prepared to accept for publication. The Internet makes widespread distribution possible to anyone with access across the Internet via a PC and modem.

The Internet has the potential to change the way that information is distributed. It is an alternative to using more traditional media and can be used to disseminate anything from an obscure personal treatise to valuable government information.

The Internet is also a useful facility to use in pursuit of the Open Government initiative. To this end the Government is keen to make available in the public interest as much official information as possible on the Internet, in line with commitments given in the Code of Practice on Access to Information. Government organisations are encouraged to make available electronic versions of paper documents on-line, as well as material which may not have been suitable for paper publication but does more usefully lend itself to electronic dissemination.

Issues such as copyright protection for content, choice of information for publication, charging and security, all need to be considered and are referred to later in this chapter. Organisations should consider carefully what they are trying to achieve from publishing information on the Internet. For example;

- advertise or promote events or publications - as an addition to free, hard copy "fliers"

- reach a wider market at negligible cost

- make use of it as an additional medium for publishing important information

- take advantage of the speed with which information can be published compared to more traditional hard copy methods.

3.2   Copyright

There is currently a fair degree of debate concerning the adequacy of existing laws to cope with publishing in the information age. Commentators range in their opinions, from those who say that copyright needs merely some adjustment, to others who feel that copyright is an anachronism.

One point of view is that computer processing or data transmission do not make any logical or legal difference to what constitutes a copy of an original work, and therefore existing laws can be applied. An opposite point of view is that copyright law is falling into disrepute as technology is making it both unenforceable and irrelevant.

Organisations which make information available on the Internet for which they have copyright, should explicitly state their copyright ownership by including appropriate notices within the body of the information (see also 3.7 of this *Guideline*). In the event of a dispute, the law will normally regard this as evidential presumption of ownership by the party which has made clear its copyright position.

If information is published (copied) onto the Internet and the publisher does not own or have sufficient rights to do so without infringing the copyright of the author of that information, then the publisher may be liable in damages for copyright infringement. Therefore, when putting information onto the Internet publishers must consider their options for acquiring rights under the existing law.

They can, for example,

- use material which is not subject to copyright protection

- use in-house developed new material made by their employees

- commission development of new material by independent contractors

- acquire rights in pre-existing material from third party copyright owners (rights clearance)

- acquire third party copyright ownership outright.

Would-be publishers need to consider carefully what types of documents to publish and whether to publish them without charge. Documents which the copyright owners might allow to be published free of charge are explanatory or promotional material, texts of speeches, etc. However, it is likely that the information most in demand will be material that an organisation might normally charge for. In the case of government information, would-be publishers are normally government organisations, working within parameters set by HMSO, which administers Crown copyright, or private sector companies licensed by HMSO.

Organisations need to establish procedures to monitor any licences under which they operate; for example, the use of software packages in an internal networked environment, is normally fairly easily controlled. However, controlling the distribution of software packages and other types of licensed information on the Internet may pose some problems. Employees should be made aware of their obligations with respect to the use and transmission of licensed material.

| 3.3 | Crown and Parliamentary copyright | HMSO is responsible for the administration of Crown copyright. Any material produced by government employees or Officers of the Crown (for example, Ministers) in the course of their duties, is automatically Crown copyright. However, if a work is commissioned by a government organisation or Crown body, copyright rests with the author or creator of the work or their |

employer, and not with the Crown, unless Crown ownership of the copyright is included specifically within the commissioning contract.

HMSO's Copyright Unit handles the practical administration of Crown and Parliamentary copyright and facilitates the widest possible use and dissemination of official material, while ensuring that all reproduction is proper and appropriate in the general public interest.

Documents subject to Crown copyright which are priced in paper form, may be allowed to be reproduced free on the Internet by a government organisation, but HMSO should be consulted. However, organisations should also consider what effect free on-line access might have on their revenue streams.

Although some government organisations have limited delegated authority to deal in copyright in some specialised material, generally speaking the HMSO Copyright Unit should always be approached for permission to reproduce any Crown and Parliamentary material.

The subject of Crown and Parliamentary copyright as well as the role of the HMSO Copyright Unit are covered in more detail at 1.16 of the *Reference Book.*

3.4   Liability

Putting information on the Internet, typically by means of a Web site, has similar consequences to publishing any other information. People may rely on it and if it is wrong and they suffer loss as a consequence, they may seek financial compensation. There is the added element that publication on the Internet constitutes publication to the world at large. Potential liability is most likely to depend on the jurisdiction in which the information is read and relied upon.

Whatever the jurisdiction, care is needed to ensure the accuracy of information (especially information which if wrong could cause personal injury or damage to property) and legal precautions should be taken to reduce the risk of liability (see 9.7 of *Reference Book*). For example, there have been several cases brought in America in which publishers have been held liable when individuals have acted on their published information and suffered loss as a result. Apparently none of these cases has succeeded, but in Europe, the publishers of a French CD-ROM on edible wild plants, have apparently been successfully sued as a result of a reader being poisoned. There has been no English case on this topic.

Where there is no likelihood of death or personal injury, the position in England is as follows. English law has traditionally been reluctant to impose liability for purely financial loss. It has tended to do so only where someone has assumed responsibility to an identifiable and small class of persons for the consequences of the advice given. Professional advisers are the obvious example. Traditionally, professional advice has been custom-made for identifiable clients but already advice such as tax tips is appearing on the Web.

If electronic and on-line publishing enables professional advice to be packaged and made available to a wider class of consumers, this could potentially provoke a change in liability for negligence towards enabling a wider class of persons to sue. Whilst this would require a considerable development of English law it may nonetheless be wise to take some precautions such as using disclaimers (see 6.2 of this *Guideline*). In addition the global, and therefore cross-border and pan-jurisdictional nature of the Internet should not be forgotten (see 10.2 of *Reference Book*).

| 3.5 | **Liability for hypertext links** | The implications of using hypertext links to other organisation's data should be made clear to all relevant personnel. For example, there may be liability implications if hypertext links refer to a defamatory statement or the possibility of criminal liability if connecting to a pornographic site. Therefore organisations with Web sites should be monitoring continuously the information they have made available on the Web and in particular their links to other sites. It is difficult, if not impossible, to avoid other organisations linking to a site, and anyway, in most cases, the link would be advantageous. |
| 3.6 | **Subscription** | Consider the use of a subscriber service requiring users to register before they can access the information. For example, Playboy uses this route in order to avoid running into trouble under the various indecency |

statutes by setting up a subscriber service whereby the subscriber needs to send various details including proof of age and/or payment. As such, the access to the site can be controlled closely.

3.7	**Disclaimers and copyright notices**	Two features which should appear on most Web sites; are disclaimers and copyright reproduction notices.

Disclaimers should be sufficiently prominent to be effective. This means that they should appear at the top of a Web page, either in full or in precis form with a clear link to where the full information can be found. Disclaimers should be drafted so as to avoid liability for reliance on the content of the Web page and to avoid liability for Web page links (see 6.2 of this *Guideline* for more information on the use of disclaimers).

Web sites should also, particularly if they contain valuable copyright information, feature a clear copyright identification to show the ownership of information in the page. They should also carry an unequivocal statement making it quite clear whether downloading in any form and onward copying are permitted or whether reproduction is prohibited or restricted. If no such clear statement is made, there is a strong likelihood that anyone accessing the material will assume that it may be freely copied (see Chapter 6 of this *Guideline*).

4 Undertaking procurement activities

4.1 Introduction

At the most basic level, the use of the Internet can assist procurement activities by speeding up and simplifying communications with potential suppliers both before and after award of contract.

Linking government organisations via the Internet can allow information, experiences and draft contract styles to be shared.

The following sections address particular aspects of the procurement process where the Internet can be used effectively, and highlight some considerations which must be appreciated prior to going on-line. Section 8.2 of the *Reference Book* provides much more detailed advice on forming an on-line contract.

Whilst the advice in this chapter is particularly relevant to public sector purchasing organisations, much of it is also appropriate to any organisation undertaking procurement activities using the Internet.

4.2 Level playing field in the public sector

Fundamentally, the UK Procurement Regulations (see 6.7 of *Reference Book* covering the procurement process are designed to ensure both the free movement of goods and services and equal conditions of competition throughout the EU. In other words, the fundamental objective is to enhance the free movement of products and services in the public purchasing sector by ensuring non-discriminatory treatment of competing suppliers. Any activity which frustrates this objective may be an infringement of the rules and may be subject to investigation and potential legal challenge.

Where a purchase involves the development of procurement documentation, such as a Statement of Requirements, e-mail may be used to minimise the effort involved in its dissemination. However, care must obviously be taken to prevent commercially sensitive information from falling into wrong hands.

To avoid an administrative law challenge, government organisations must ensure that all tenderers receive the same amount of information so that they are able to tender from a "level playing field"; that is all competitors should have the same chance to win the contract. Therefore, if one supplier has an e-mail link with a government organisation, the same opportunity should be given to all the suppliers. However, at present, this should not be made into a project requirement.

4.3 Use of EDI

Principles for the use of EDI have been formulated in various sectors of industry and in the public sector and are used successfully (see 8.5.1 of *Reference Book*). Although there are different EDI standards across different industry sectors, those standards have been formalised and used for a number of years.

Generally, the higher the value of the contract the greater the need for certainty of contract terms. In the public sector, where procurement is subject to the regulations, Departments are under the scrutiny of statutory auditing bodies and need to make certain that every decision and communication during the procurement process is justified and documented in order to accord with best practice. Such considerations highlight the benefits of using the structured requirements of EDI.

4.4 Need for hard copy records

The requirements of audit and for clear evidence, combine with the requirements of statute, to compel public sector organisations involved in procurement activities to maintain suitable records (see Chapter 4 of *Reference Book*).

Electronic records present opportunities to reduce the physical limitations of paper storage. However, in order to use electronic records successfully, clear internal guidelines should be in place in order to establish the validity and legal weight of any electronic document which finds itself presented before a court of law if a contract is litigated (See 7.4 and 7.5 of *Reference Book*). At present, it is recommended that contracts of any significance should be executed in hard-copy format in order to avoid difficult evidential issues if the content is subject to litigation.

It is very important to ensure all computer data is backed up successfully because any loss of data may result in a failure to comply with duties imposed by statute.

4.5 **Publication of Official Journal advertisements**

The UK Procurement Regulations include rules governing the advertising of procurement requirements by central government organisations and "bodies governed by public law". The value of the supplies and/or services required determines whether an advertisement needs to be placed in the Official Journal of the European Communities.

The Internet is recommended as a method for the publication of notices in the Official Journal. To stay within the UK Procurement Regulations, the notice must not appear on the Internet before the date when the notice is dispatched to the Official Journal and must not contain any additional information to the Official Journal notice.

The SIMAP project (in English, the Public Procurement Information Systems Project) involves the use of EDI and is being undertaken under the auspices of the Commission's Interchange of Data between Administrations (IDA) programme.

The pilot stage of the project commenced in November 1995 and runs for 12 months. Dependent upon the outcome of the trial, the Commission is planning to adopt the SIMAP system as the normal means for transmitting notices for publication in the Official Journal.

CCTA is continuing to participate in the pilot project. A particular concern is to ensure accuracy and legal compliance with UK Regulations for notices sent for publication in the Official Journal.

5 Connecting to and using the Internet

5.1 Introduction

Chapter 1 provides guidance on the responsible use of e-mail; one of the most commonly used facilities on the Internet. This chapter looks at some of the other facilities available on the Internet and provides guidance on their responsible use. To assist in understanding the terminology, this introductory section gives a brief description of some of the more common terms used in connection with the Internet.

Essentially, the Internet consists of thousands of individual computers and networks talking to each other using a common language or **protocol**.

The usage and influence of the Internet has broadened considerably from its military and academic roots and Internet protocols have been adopted by many organisations. Each network on the Internet is linked either directly by telecommunications links or, more likely, via an **Internet Service Provider**. The service providers are responsible for the links of their part of the system; no one organisation "owns" the Internet. Technical standards are issued and updated by committee, and bodies in various countries are responsible for giving each computer which becomes part of the Internet a unique name so that it can be addressed by any other computer.

The Internet has traditionally been used for e-mail, discussion forums (known as **Usenet newsgroups**) and for the transfer of files. However, it was far from easy for the non-technically minded to use. The recent explosion has come about because of the newest face of the Internet; the **World Wide Web (the Web)**.

The Web is not a separate part of the Internet but a user-friendly way of searching for or **surfing** information contained on the Internet. The Web allows organisations to set up a **home page**, which is usually a page of text and/or graphics introducing the surfer to the organisation. Within the page, certain highlighted words or icons are "linked" to other pages on the same **Web site** or to other Web sites, perhaps on another computer in another part of the world. Simply by clicking with a mouse on these **hypertext links**, the surfer can jump to the other information. The page is written in a format which supports these links, known as **Hypertext Mark-Up Language**, or **HTML**.

To access the Web, a user requires an account with an Internet service provider. Some providers simply allow access through their computers, by dial-up line and a modem, or by leased line. Some, in addition, are information providers themselves. Each provides the user with the appropriate software, known as a **Web browser** (or search engine), to allow the user's computer to receive and display the Web information and the coded hypertext links. Other software is required to access newsgroups, access files and to send e-mails.

The Web is a multimedia environment; it is possible to access Web sites which contain video and sound clips as well as text and graphics. The main problem at the moment is lack of **bandwidth**, which is a measure of the amount of data which can be sent over a telecommunications link at one time. Video and audio require wide bandwidth, but many users access the Internet over normal telephone lines which can handle only limited bandwidth. The situation is likely to improve as investment in the infrastructure of the Internet increases.

Other terms which are used in connection with the Internet and the Web include:-

Domain name

A domain name is a unique name given to a particular Internet site, usually read from right to left, and consists of two or more alphanumeric fields, separated by a "dot". Each field is called a sub-domain. In the UK the rightmost sub-domain, called the top-level domain, is geographical denoting the country the computer is located in. The next sub-domain indicates the type of organisation, whether it is, for example, a government (gov), commercial (co) or academic (ac) organisation; the third sub-domain is the name of the organisation, and the leftmost sub-domain is the name of a specific computer. There can be a further sub-domain denoting, for example, a department within the organisation, which would come between the type of organisation and the name of the organisation (this is not shown in the following example):

[computer].[organisation name].[type of organisation].[country code]

mc2.thecompany.co.uk

e-mail address

Each person's e-mail address is unique and consists of a user name called a user-id, which may be numeric in some cases, and a domain name separated by the symbol "@"; for example:

a.student@newcastle.ac.uk
765432.123@compuserve.com

In the UK the fields in the domain name part of the address are sometimes in reverse order; for example:

a.student@uk.ac.newcastle

Uniform Resource Locator (URL)

The address of a Web page or a pointer to information on the Web which can include pointers to resources such as ftp servers as well as Web servers. The protocol for the Web is called the hypertext protocol or **http**, and each address usually begins "http://www" followed by the domain names, for example:

http:/www.open.gov.uk

Much of the information available over the Internet is held on university computers, and is managed by individuals. Similarly, many of the Internet indexing and cataloguing services available (for example, Archie and Gopher) have been set up and run by University departments.

Each Internet provider offers specific Internet services, and different providers offer different packages. The main services are:

- e-mail

- newsgroups; for example, the Usenet with approximately 10,000 discussion groups

- FAQs - Frequently Asked Questions files; these are collections of the most commonly asked questions and answers regarding a particular subject

- Discussion List or Listserv - similar to the Usenet newsgroups except that information is sent and received through e-mail

- Gopher, Archie, etc. - search tools for the Internet

- File Transfer Protocol (FTP) - a program and protocol for transferring files between machines

- Telnet - a protocol that enables a user to log into a remote machine

- World Wide Web (WWW) - the hypertext document system.

Another new facet of the Internet is Internet telephony. This is the ability to hold real-time voice telephony conversations, with the bonus of local call charges. There are, however, regulations governing Public Telephone Operators (PTOs) whereby they need to have a licence before they can offer a service (see 6.4 of *Reference Book*).

| 5.2 | **Security and confidentiality** |

Encryption has already been referred to at 2.5 but, clearly, its use can help fight the threats of corruption to data and unauthorised access.

Organisations need to consider the protection of their sensitive information. Employing security measures helps to ensure privacy (see 9.9.1 of *Reference Book*).

Additionally, many organisations employ "firewalls"; a collection of hardware and software components that together provide a protective channel between the internal network and the Internet. A firewall can filter and check the information before it is allowed into the internal network. It can also allow restricted access to certain Internet port numbers and block access to everything else. However, firewalls are not 100% secure and some organisations may feel compelled to set up a standalone PC or network, separate from the organisation's main network (an "air gap"). It is a matter of the degree of risk that the organisation wants to accept (see 9.9.2 of *Reference Book*).

Legal mechanisms are also available either to prevent disclosure of confidential information, or once a disclosure has occurred, to recover any financial loss which may result. Confidential information can be protected either by contract or through an action for breach of confidence.

5.3	**Newsgroup usage**	It should be borne in mind that newsgroups are intended to be public and the postings are in effect published to the world at large; in other words they are analogous to speaking in a public forum. The security aspects of newsgroup usage are referred to at 9.9.1 of the *Reference Book*.
5.3.1	Reading Usenet newsgroups	It is difficult to stop access to Usenet groups unless specific blocking software is used. Organisations may feel compelled to forbid downloads, or to allow only downloads to a "safe area" where virus introduction can be controlled.
5.3.2	Posting to Usenet newsgroups	Organisations may wish to place restrictions on which staff may post to Usenet newsgroups. Postings must not contain material that is:

- illegal or prohibited by any law or regulation in any jurisdiction, or

- could infringe any rights of, or render the sender liable to, any person.

In certain circumstances it may be appropriate to forward the content of a posting to an in-house lawyer for approval prior to it being sent.

5.3.3	Personal postings	Organisations may consider banning personal postings to Usenet newsgroups (although this may be difficult in practice). An alternative approach is to require employees to include the following statement at the end of any personal posting:

"This posting is a personal communication and is not authorised by or sent on behalf of any other person or company."

5.4   Web page access

The passive nature of Web pages means that persons within organisations who access a Web page are less of a threat than persons using e-mail or the Usenet. However, the existence of response forms and/or mailback facilities at Web sites can pose potential problems for organisations. For example, Web pages frequently present opportunities to the reader to form contracts, and employees accessing Web sites must have clear guidelines as to what is permitted (see 8.2.1 of *Reference Book*).

Security concerns regarding downloadable files apply equally to files downloaded from Web sites. Generally, no file should be downloaded unless doing so is expressly permitted by the terms of the Web site. Failure to observe such rules can expose organisations to serious civil and criminal penalties. Additionally, organisations need clear procedures concerning downloading files so as to avoid problems such as the importation of viruses and other malicious code, pornography, and copyright protected material.

Allowing unregulated Web usage by employees can put organisations at risk of prosecution. In 1994, a multinational company in the UK was investigated by Scotland Yard's Obscene Publications Squad because a young employee had stored some forty obscene images in the company's computer. The company was unaware of the employee's activities but if organisations are judged to in any way have conspired or abetted in an employee's actions then they are liable for prosecution. Not having procedures to safeguard against an incident like this might be construed as abetting.

6   Setting up a Web page

6.1   Introduction

In technical terms, the Web is a wide-area, hypermedia, information retrieval initiative aiming to give universal access to a large number of documents. In everyday terms, the Web is a user-friendly way of searching for information on the Internet.

Before products or services can be marketed or sold the seller needs to establish a Web site or home page as its virtual shop window. The operation of the Web relies on hypertext as its means of interacting with users.

Hypertext connects documents. It is basically the same as normal text on a page except that there are Hypertext Mark-up Language (HTML) links to other documents. These documents are often held on other servers, and in other countries.

Hypertext links invariably appear in a different colour to the regular text. When the cursor is allowed to rest on the link the Uniform Resource Locator (URL) appears showing the location of the information referred to. By using HTML, any page can link to any number of other pages to create a complex virtual web of connections. Such "open-plan" architecture makes it very important to decide the boundaries of the organisation's liability for the content it puts on the Web. The lines can be drawn using disclaimers, clear internal guidelines and user awareness programmes which highlight the potential risks.

Hypermedia is the technological extension to hypertext which allows links to be made to not only other pieces of text but also other forms of media such as sounds, images and video.

Documents on the Web are referred to using URLs.
A simple URL may look like:

http://www.twobirds.co.uk/*name-of-document*.html.

It consists of three parts:

- the method of retrieving the document (http)

- the name of the domain where the documents are held (www.twobirds.co.uk)

- the path name to the document (*name-of-document*.html).

To hold information on the Web requires a server such as an HTTP, Gopher, FTP or WAIS database. Whichever server is chosen depends on individual needs, but, in general, if the requirement is to deliver hypertext with reasonable speed then an HTTP server is used.

6.2   Use of disclaimers

A disclaimer is a formal notice that seeks to allow the party making the disclaimer to avoid liability to others under certain circumstances. For example, organisations in the services sector whose business is giving advice in a certain sphere may wish to disclaim liability for loss suffered by a third party for whom the advice given on the Web site was not appropriate.

Disclaimers often take the following format:

"This Web site contains general information concerning (xx organisation). Nothing on this site is to be taken as constituting specific (for example, financial) *advice. Always consult a suitably qualified* (for example, accountant) *on any specific matter"*.

The extent to which disclaimers can restrict or limit liability varies from country to country; clearly a problem given the global nature of the Internet. It is advisable to create a situation where a Web site is expressly governed by the law of a particular country; for example, by stating that the law of England and Wales applies to the content of the Web site. HTML links complicate this ideal and the rules governing which country's laws apply are very complex.

The effectiveness of a disclaimer varies depending upon which country's courts are testing the language of the disclaimer. It is clear, however, that a suitably worded disclaimer should disclaim liability in most situations.

In addition to the fact that foreign laws have a different view of an "English" disclaimer, even in the UK there are certain liabilities which cannot be disclaimed against. For example, liability for defamation cannot be effectively disclaimed.

Disclaimers should attempt to avoid liability for loss suffered by anyone relying upon that information. It should be borne in mind that if information placed upon a Web page is of a general nature, liability arising from its use is less likely than if the information is of a specific nature.

Many users expect information on government organisation's Web sites to be completely up to date but this is not always the case, therefore a disclaimer to that effect about the accuracy or currency of the material may be needed.

Many countries (including the UK) restrict the ability of a party to avoid some kinds of liability. For example, such laws exist particularly to restrict those who would seek to avoid liability for death or personal injury caused by reliance upon their Web site content.

The use of contractual provisions to exclude or limit liability is regulated by the Unfair Contract Terms Act 1977 and the Unfair Terms in Consumer Contracts Regulations 1994. The 1977 Act imposes a requirement that any term which attempts to exclude or limit liability be "reasonable", assessed according to a range of factors. The 1994 Regulations invalidate any terms which are considered "unfair".

A disclaimer should be drawn to the user's attention by, for example, the use of bold type. A good approach is to have the disclaimer as the first item visible upon accessing a Web page. On some Web sites it is possible to have disclaimers that remain on-screen as the user scrolls through the information that is being presented. This is particularly valuable in circumstances where users may go directly to a specific Web page on a Web site and as a consequence fail to be presented with any disclaimers set at the Web site level.

6.3   HTML links

The Web allows many pieces of information on many computers in numerous countries to be linked together. The down side is that a Web page over which control is exercised may become linked to a page which is less conscientiously maintained.

Web page owners need to consider whether linking to another page can make them responsible for what is on the other pages. Whilst there is currently no precedent in English law in relation to the Internet, it is possible that a page owner can be sued if there is a link to a libellous page (see 2.5.3 of *Reference Book*). There is ancient law which states that a party who draws attention to a libel may also become liable in damages to the person defamed. It is clear that the use of HTML links may do just that.

Good practice requires decisions be made as to whether to link a site to others and, from a wider perspective, who is charged with keeping control of a site. It is difficult to restrict incoming links from external sites. Indeed, in many instances such links can be beneficial in both marketing and awareness terms.

Users may not recognise immediately what is, and what is not, information belonging to a particular site. Therefore, it is wise to state clearly which information relates to a particular organisation in order not to assume responsibility for third party material.

6.4	**Interactivity**	A normal Web site is analogous to an advertising hoarding. Further, a Web page may be enhanced by use of response or feedback mechanisms and/or ordering facilities. However, there is no point in having the most fascinating site on the Web if it is always so busy that potential visitors cannot gain access. Similarly, Internet users expect to be able to access and retrieve information quickly, therefore the perceived level of traffic is an important factor in the choice of Internet provider.
6.5	**Liability for content**	As mentioned at 6.2 above, it is clear that organisations need to take effective steps to limit liability for the content of their Web sites by the effective use of disclaimers.

A corollary to the message in 6.2 above is that organisations should take active steps to update and monitor their Web sites. Organisations should exercise appropriate control over content and are likely to be required to indemnify the Internet service provider should there be a claim; for example, for copyright infringement. Therefore, organisations should either own the rights in the content held on their sites or secure effective rights to include third party material.

Organisations should implement appropriate procedures to make sure that reasonable care is taken, as required by the DPA 1984, to ensure that personal information placed on a Web page is accurate (see 3.3.6 of *Reference Book*). If a recipient suffers loss from relying on inaccurate information provided by the sender, and the sender was, or should have been, aware that such reliance would occur, then a court may find the sender liable in negligence for such misinformation. Where data accuracy cannot be reasonably assured; for example, because the information was provided by a third party, this should be brought to the attention of the recipient by the use of a disclaimer.

Glossary

air gap	Where an internal network is kept physically separate from the Internet.
Archie	A system for locating files that are publicly available on the Internet.
cyber bucks	A euphemism for electronic currency on the Internet.
EDI	Electronic Data Interchange. The transfer, from computer to computer, of data using an agreed standard to structure the message content.
e-mail	Messages sent by a user over networks to other users.
EU	European Union.
FTP	File transfer protocol. A protocol that defines how to transfer files from one computer to another. It can also mean the program which serves the files using the protocol.
Gopher	A look-up tool for searching the Internet and selecting resources (such as files or other menus) from menus. It is like browsing through a remote library's card catalogue.
Hacker	A common term meaning a person who intentionally seeks to obtain unauthorised access to computer programs or data.
HMSO	Her Majesty's Stationery Office.
HTML	HyperText Mark-up Language. The language in which World Wide Web documents, often referred to as pages, are written and which includes computer instructions on displaying the information on the pages.
HTTP	HyperText Transfer Protocol. The network protocol used for transferring data on the World Wide Web.
Hypertext	Text that contains electronic links to other text, files, pictures etc. within the same document or other documents. Selecting a word or phrase (which is usually highlighted) will automatically provide other information about the word or phrase.

Internet	Technically it is a global "network of networks" connected to each other using common protocols. However it is also a vast source of information in different forms and a way in which people worldwide can communicate with each other.
Internet port number	A number that identifies a particular Internet application.
Internet service provider	An organisation which provides facilities to access the Internet.
IT	Information Technology.
modem	A piece of equipment which connects a computer to a data transmission line (typically a telephone line).
PC	Personal computer.
pkc	Public key cryptography. A powerful technique for encrypting, hashing and signing messages or files.
server	The computer and associated software which offers a service to another computer; for example in the case of a file server, providing a file that is requested.
URL	Uniform Resource Locator. A pointer to where a resource (such as a document) is located on the Internet.
Usenet	A large collection of discussion groups involving millions of people. Each discussion group usually centres around a particular topic.
vicarious liability	Liability which falls on one person as a result of an action or crime of another; for example the liability of employers for the acts and omissions of their employees.
WAIS	Wide Area Information Server. A powerful system for looking up indexed information in databases (or libraries) on the Internet.
World Wide Web	Often known as the Web. A hypertext based system for finding and accessing information on the Internet.

Index

To help the reader, this index covers both the *Guideline* and the *Reference Book*. References to the *Guideline* are preceded by the letter 'G' and references to the *Reference Book* are preceded by the letter 'R'. Where an item has references to both books the 'R' references come first.

Printed in the United Kingdom for HMSO
Dd302583 5/96 C10 G3397 10170

Legal issues
and the Internet

Reference Book

LEEDS LIBRARY & INFORMATION SERVICES
DISCARDED

NOTE: Any references to the law are current at the date
of issue of this Reference publication

LONDON: HMSO

CCTA
Central Computer and Telecommunications Agency

LD 0999979 5

Acknowledgements
The assistance of Bird & Bird, and in particular Mark O'Conor, in the preparation of this volume under contract to CCTA is gratefully acknowledged.

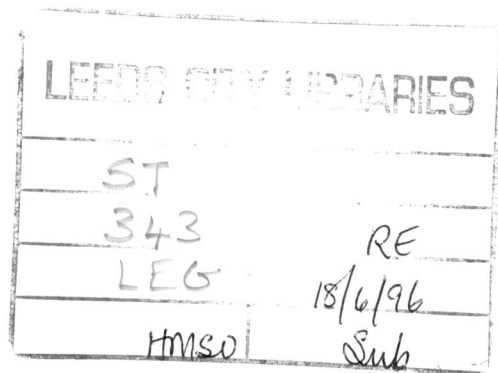

LEEDS CITY LIBRARIES

ST
343
LEG
HMSO

RE
18/6/96
Sub

© **Crown Copyright 1996**
Application for reproduction should be made to HMSO Copyright Unit

ISBN 0 11 330682 2

For further information regarding this publication and other CCTA products please contact:

CCTA Library
Rosebery Court
St Andrews Business Park
Norwich NR7 0HS
01603 704930

Contents

Management Summary

Use of the Internet to communicate, buy and sell and publish, presents interesting but important paradoxes. On the one hand the truly global and open nature of the Internet enables people to access it from wherever and whenever they like. On the other hand the onus is on people to take the responsibility for their actions from a legal, moral and security standpoint.

Using the Internet in all its forms and guises is very straightforward but making sure that what has been done is safe from a myriad of directions, is both complex and a veritable minefield.

This *Reference Book*, which can be read in its entirety or referenced through the table of contents and the subject index, provides an in-depth analysis of the relevant laws and legal issues relating to the use of the Internet. It expands on the companion *Guideline* which examines some of the main ways in which business uses the Internet and the legal issues involved.

This *Reference Book*, through an analysis of the relevant principal laws of England and Wales, identifies the implications of the law in a straightforward manner that the Internet user can understand and act upon. Although the majority of the references to the law in this publication apply to the United Kingdom as a whole, the law in Scotland and Northern Ireland does have subtle differences to the law in England and Wales. This means that anyone applying the information given in this publication, within Scotland or Northern Ireland, will need to bear this in mind.

Users of the Internet in general, and management in particular, need to appreciate that its use needs to be supervised. There is a need for control and a conscious effort made to identify, understand and assess the risks involved.

1 Copyright and other intellectual property rights

1.1 Introduction

The terms intellectual property and intellectual property rights (IPR) describe a bundle of rights encompassing copyright, patent, trade marks, service marks, design rights, trade secrets and other similar proprietary rights. Being a property right, any of the above may be dealt with in a way analogous to dealing with property; that is they may be sold outright (parting with ownership) or licensed to another party (merely parting with possession but retaining ownership). Copyright is a subset of the panoply of IPR.

Copyright is the authors' right to control and regulate use by others of their original works. It gives authors the right to seek financial returns for distribution of their works and encourages further creative effort. Copyright enables copyright owners to determine the types of use to which their works are put. It gives authors and publishers the security to exploit their works and it creates legal remedies for unlawful copying in the market place.

The "information society", as its name suggests, is concerned with information. In order to encourage investment into new digital technology, the owners and controllers of such information need to know that it is secure (see also Chapter 9 on Security) and that their rights are protected. Because of the absence of direct contractual links between the majority of parties using the Internet, copyright assumes a high profile as an important method of protection.

The World Wide Web (Web) provides an opportunity for organisations to distribute information electronically by, for example, providing an on-line version of a magazine, or exhibiting examples of their own publications. However, there is a need to ensure that rights are protected. Conversely, organisations using the Web as a source of information need to ensure that the IPR of the sender of the information, or some other third party owner, are not infringed, thereby exposing the organisation to potential civil and criminal liability.

When potentially copyrightable material is produced in electronic form, it is capable of being easily transmitted, reproduced and modified. This presents significant challenges for existing IPR. Many countries are considering reform of their national laws and some are even undertaking a consultation process. The European Commission published in 1995 a Green Paper entitled "Copyright and Related Rights in the Information Society" which requested views from interested parties on the sufficiency and future of copyright law[1]. The Green Paper also made various suggestions for the development of copyright law at the Community level. There is general debate in the media and between practitioners as to whether copyright law must change to cope with the realities of the global information society.

Modern copyright law is set out in the Copyright, Designs and Patents Act 1988 (CDPA). This attempts to balance the needs of users to access and exchange information, with the rights to financial reward of authors and publishers.

Traditionally, copyright is applied to works on a media by media basis. Holding such works in electronic form may make these distinctions redundant.

1.2 Some basic facts

1.2.1 Sources of law

The current statute, the CDPA:

"restates and amends the law of copyright"
[CDPA s.172(I)].

The CDPA had, as one of its stated aims, the use of everyday language as far as possible. It explains that provisions under previous law may be referred to in order to help interpret its sections [see CDPA s.172(3)].

The CDPA has been amended by the Copyright (Computer Programs) Regulations 1992 (the UK Software Regulations), which took effect on 1 January 1993, and the Duration of Copyright and Rights in Performances Regulations 1995 (SI1995/3297), which came into effect on 1 January 1996.

---

[1] COM (95) 382 final

1.2.2   Subject matter

CDPA s.1(1)(a) states that:

"Copyright is a property right which subsists in original literary works".

Copyright also subsists in:

- original dramatic, musical or artistic works

- sound recordings, films, broadcasts or cable programmes

- the typographical arrangement of a published edition.

Compilations, tables and computer programs are included as literary works, ("compilations" and "computer programs" are undefined).

Films include video recordings and any moving images. Cable programmes include publicly accessible on-line databases.

1.2.3   Originality

There is no definition of "originality" in the CDPA. The EU Software Directive in Article 1(3) defines "original" to mean "the author's own intellectual creation" but this definition is not included in the UK Software Regulations.

Nevertheless a work must be original to be regarded as copyright. Generally, mere facts are regarded as non-copyright.

1.2.4   Fixation

What separates an uncopyrightable idea and a copyrightable expression of that idea is the fact that in the latter case the idea has been expressed in some way; that is, fixed in some form. This fixation can be by means of writing down in the case of a literary work which is stated in CDPA s.3(2) as follows:

"Copyright does not subsist in a literary ... work unless and until it is recorded, in writing or otherwise".

In the definition section of the CDPA, s.178 states that "writing" includes:

"any form of notation or code, whether by hand or otherwise and regardless of the method by which, or medium in or on which, it is recorded".

It is clear then that the contents of a screen are copyrightable (subject to its meeting the other criteria, such as originality). This is notwithstanding the fact that the display is the product of the internal workings of a computer, and not "writing" in the sense of being handwritten.

On the Internet, material and information is accessed and (temporarily at least) resides in the computer's RAM. In order to see that information on screen, it has been displayed by the computer (see 1.6.1 as to how such activity can constitute infringement of the rights of a copyright owner).

1.2.5   Idea versus expression

Although not expressly stated in the CDPA, it is generally accepted that copyright protects the expression of ideas but not ideas themselves (for example, *John Richardson Computers Limited v Flanders & Chemtec Ltd* [1993] FSR 497, at 523).

Compare this to the EU Software Directive Article 1(2), which states:

"Protection in accordance with this Directive shall apply to the expression in any form of a computer program. Ideas and principles which underlie any element of the computer program, including those which underlie its interfaces, are not protected by copyright under this Directive".

This provision of the EC Software Directive was not expressly incorporated in the UK Software Regulations on the ground that it was already part of judge-made copyright law in the UK. Thus, it is possible to speak of an "idea/expression dichotomy" in UK copyright law akin to the "idea/expression dichotomy" of US law.

1.3   Authorship and first ownership

The CDPA contains the following definitions regarding authorship and ownership of copyright:

Individual authorship

An *"author"*, in relation to a work, means *"the person who creates it"* [CDPA s.9(1)].

Joint authorship

A *"work of joint authorship"* means:

"a work produced by the collaboration of two or more authors in which the contribution of each author is not distinct from that of the other author or authors" [CDPA s.10(1)].

References to the *"author of a work"* in the CDPA:

"shall, except as otherwise provided, be construed in relation to a work of joint authorship as references to all the authors of the work" [CDPA s.10(3)].

Authorship of computer-generated works

CDPA s. 9(3) states that:

"In the case of a literary ... work which is computer-generated, the author shall be taken to be the person by whom the arrangements necessary for the creation of the work are undertaken".

CDPA s.178 states the term "computer-generated" means that:

"the work is generated by computer in circumstances such that there is no human author of the work".

First ownership in general

CDPA s.11(1) states, subject to certain provisions relating to works made in employment (see below) and Crown and Parliamentary copyright (see 1.16), that:

"The author of a work is the first owner of any copyright in it..."

Works made in employment

CDPA s.11(2) states that:

"Where a literary ... work is made by an employee in the course of his employment, his employer is the first owner of any copyright in the work subject to any agreement to the contrary".

But see *John Richardson Computers Limited v Flanders & Chemtec Ltd* [1993] FSR 497, at 516-19 ruling that copyright in the plaintiff's program was held by the defendant upon an implied trust for the plaintiff despite the absence of an assignment from the defendant and despite the fact that for some of the time the defendant, when writing the plaintiff's program, was engaged to do so as an independent contractor rather than as an employee.

Under normal principles, the defendant would have owned the copyright in the program himself, being an independent contractor, but in the above case the circumstances were such that the copyright was said to be held by the plaintiff. The implied trust mechanism was employed by the trial judge because normally a copyright cannot be passed from one party to another without an assignment in writing; that is, an executed document stating that the copyright is to pass from one party to another. The implied trust means that although technically the defendant holds the copyright in the program, he holds it for the benefit of the plaintiff.

It is important therefore to ensure that appropriate mechanisms are in place to deal with the ownership of the rights in content (see 1.18.4).

1.4   Duration

1.4.1 Term of protection

Prior to 1 January 1996 copyright in original literary, dramatic, musical or artistic works lasted for 50 years beyond the year in which the author died. If there were joint authors then copyright duration was linked to the last death of any identifiable author.

Copyright in sound recordings, films, broadcasts or cable programmes lasts for 50 years from the year in which the work was first made or released or broadcast.

Copyright in the typographical arrangement of a published edition lasts for 25 years from the year of first publication.

1.4.2 Extension of term

Under the recent EU Directive (93/98/EEC) harmonising the term of protection of copyright and related rights, the term of copyright protection in the UK for literary works - including compilations - has been extended to the lifetime of the author plus 70 years. This Directive has been implemented in the UK by the Duration of Copyright and Rights in Performances Regulations 1995 (SI1995/3297), which came into effect on 1 January 1996.

1.5   National origin

Copyright does not subsist in a work unless the author is a qualifying person or the work was first published in a qualifying country (see CDPA s.153).

A qualifying person includes a British citizen, an individual domiciled or resident in the UK or another country to which the Act has been extended (including the Member States of the Berne Convention and the Universal Copyright Convention), or a body incorporated under the law of the United Kingdom or another country to which the Act has been extended (see CDPA s.154).

A qualifying country includes the UK or another country to which the Act has been extended (including the Member States of the Berne Convention and the Universal Copyright Convention) (see CDPA s.155).

1.6   Restricted acts

1.6.1   Primary infringement

CDPA s.16(1) establishes that:

"the owner of the copyright in a work has ... the exclusive right to do the following acts in the United Kingdom –

(a) to copy the work (see section 17)

(b) to issue copies of the work to the public (see section 18)

(c) to perform, show or play the work in public (see section 19)

(d) to broadcast the work or include it in a cable programme service (see section 20)

(e) to make an adaptation of the work or do any of the above in relation to an adaptation (see section 21)".

If anyone infringes this right, liability can arise whether or not the infringing party is aware of such infringement.

Copying and computer storage

CDPA s.17(2) states that:

"Copying in relation to a literary, dramatic, musical or artistic work means reproducing the work in any material form. This includes storing the work in any medium by electronic means".

Incidental copying

CDPA s.17(6) states that:

"Copying in relation to any description of work includes the making of copies which are transient or are incidental to some other use of the work".

Distribution and rental right

As originally enacted, the CDPA created a distribution right which included a new rental right in sound recordings, films and computer programs [see s.18(1)]. However the distribution right (but not the rental right) was subject to a worldwide "exhaustion" doctrine;

whereby the first sale of a copy of the work anywhere in the world exhausted the copyright owner's right to control the subsequent sale of that copy [see s.18(2)].

The rental right in sound recordings, films and computer programs was not subject to the exhaustion doctrine; therefore the copyright owner could control the rental of a copy of those works notwithstanding the first sale of that copy anywhere in the world.

The EU Software Directive called for the creation of a distribution right, including a rental right, in computer programs, and provided that the distribution right (but not the rental right) should be subject to a Community-wide exhaustion doctrine [see Article 4(c)].

The UK Software Regulations (see Regulation 4) have implemented the EC Software Directive by retaining the rental right in computer programs, providing that the distribution right is subject to a Community-wide rather than a worldwide exhaustion doctrine. This means that the first sale of a copy of a computer program within the Community, but not elsewhere, exhausts the copyright owner's right to control the subsequent distribution (but not rental) of that copy.

CDPA s.178 states that a *"rental"* means:

"any arrangement under which a copy of a work is made available –

> *(a) for payment (in money or money's worth) or*

> *(b) in the course of a business, as part of services or amenities for which payment is made, on terms that it will or may be returned".*

Furthermore, the rental right in computer programs applies to rentals by public libraries in the UK, whether or not a charge is made [see CDPA Schedule 7, paragraph 6 (Scotland), paragraph 8 (England and Wales)].

Adaptation and program conversion

Section 21(3) (a) and (b) of the CDPA as amended by the UK Software Regulations [see Regulation 5(3)], now provides that *"adaptation"* in relation to a computer program means:

"an arrangement or altered version of the program or a translation of it".

Section 21(4) of the CDPA as amended by the UK Software Regulations [see Regulation 5(3)], goes on to provide that a "translation" in relation to a computer program:

"includes a version of the program in which it is converted into or out of a computer language or code or into a different computer language or code".

1.6.2 Secondary infringement   The CDPA establishes five "secondary infringements":

- importing an infringing copy (s.22)

- possessing or dealing with an infringing copy (s.23)

- providing the means for making infringing copies (s.24)

- permitting the use of premises for infringing performance (s.25)

- providing the apparatus for an infringing performance (s.26).

An "infringing copy" is defined in s.27 of the CDPA. In addition, as amended by the UK Software Regulations, s.27(3A) of the CDPA now provides a revised definition of an "infringing copy" of a computer program. Liability for secondary infringement is different to primary infringement in that it requires that the infringer either knows, or has reason to believe, that such an infringement has been committed.

Electronic transmission   CDPA s.24(2) states that:

"Copyright in a work is infringed by a person who without the licence of the copyright owner transmits the work by means of a telecommunications system (otherwise than by broadcasting or inclusion in a cable programme service) knowing or having reason to believe that infringing copies of the work will be made by means of the reception of the transmission in the United Kingdom or elsewhere"

Putting copyright works on the Internet may therefore infringe the rights of the copyright owner, just as may a person who actually transmits that work. However, there is some uncertainty as to whether passive distribution, for example by a bulletin board operator, would be seen by a court as providing the means for making infringing copies.

1.6.3 Devices designed to circumvent copy-protection

CDPA s.296(1) states that the Act applies:

"where copies of a copyright work are issued to the public, by or with the licence of the copyright owner, in an electronic form which is copy-protected".

CDPA s.296(4) states that references to copy-protection include:

"any device or means intended to prevent or restrict copying of a work or to impair the quality of copies made".

CDPA s.296(2) states that:

"The person issuing the copies to the public has the same rights against the person who, knowing or having reason to believe that it will be used to make infringing copies –

> *(a) makes, imports, sells, or lets for hire, offers or exposes for sale or hire, or advertises for sale or hire, any device or means specifically designed or adapted to circumvent the form of copy-protection employed, or*

> *(b) publishes information intended to enable or assist persons to circumvent that form of copy-protection,*

as a copyright owner has in respect of an infringement of copyright".

Following CDPA s.296(2), the UK Software Regulations expands the scope of infringing acts by adding a new sub-section (2A).

1.7 Exceptions to the restricted acts

1.7.1 Research and private study

CDPA s.29(1) states that:

"Fair dealing with a literary, dramatic, musical or artistic work for the purposes of research or private study does not infringe any copyright in the work".

The UK Software Regulations add a new subsection (4) to CDPA s.29 which excludes decompilation from the scope of the fair dealing defence.

1.7.2	Back-up copying	The UK Software regulations have introduced a new s.50A to the CDPA permitting the making of back-up copies under certain circumstances. This new section allows a lawful user of a program to make any necessary back-up. This right cannot be excluded by contract.
1.7.3	Decompilation	The UK Software Regulations have introduced a new s.50B to the CDPA which permits "decompilation" under certain circumstances. In essence, those circumstances are that decompilation is required to achieve interoperability.
1.7.4	Other permitted acts	The UK Software Regulations (Regulation 8) provide that certain other acts of copying or adaptation (including error correction) are permitted if not prohibited by contract.
1.8	**Moral rights**	The CDPA identifies four elements to moral rights which subsist alongside copyright:

a) **The paternity right**; the right to be identified as author (or director of a film) [see CDPA s.77]. This right does not exist unless it is asserted by the author or film maker

b) **The integrity right**; the right to object to the derogatory treatment of a work [see CDPA s.80]

c) **False attribution**; the right not to suffer false attribution of a work. This lasts for life plus 20 years only [see CDPA s.84]

d) **Privacy of photographs**; the right of a person commissioning photographs not to have copies issued to the public [see CDPA s.85].

Moral rights cannot be assigned, but they may be waived. Infringement of the above rights is actionable as a breach of a statutory duty. The new moral rights of paternity and integrity do not apply to computer programs or computer-generated works [see CDPA s.79(2) and s.81(2)].

The existing right to object to false attribution of a work does apply to such works [see CDPA s.84].

| 1.9 | **Avoidance of certain contract terms** | The UK Software Regulations (Regulation 11) have introduced a new s.296A to the CDPA which provides that a contract term is void insofar as it prohibits the right to make back-up copies, decompile and: |

"observe, study or test the functioning of the program in order to understand the ideas or principles which underlie any element of the program".

| 1.10 | **Transfers of copies of works in electronic form** | CDPA s.56(1) states that the Act applies: |

"where a copy of a work in electronic form has been purchased on terms which, expressly or impliedly or by virtue of any rule of law, allow the purchaser to copy the work, or to adapt it or make copies of an adaptation, in connection with his use of it".

CDPA s. 56(2) states:

"If there are no express terms:

> *a) prohibiting the transfer of the copy by the purchaser, imposing obligations which continue after a transfer, prohibiting the assignment of any licence or terminating any licence on a transfer, or*

> *b) providing for the terms on which a transferee may do the things which the purchaser was permitted to do,*

anything which the purchaser was allowed to do may also be done without infringement of copyright by a transferee; but any copy, adaptation or a copy of an adaptation made by the purchaser which is not also transferred shall be treated as an infringing copy for all purposes after the transfer".

The provisions also apply:

"where the original purchased copy is no longer usable and what is transferred is a further copy used in its place"
[see CDPA s.56(3)] and

"on a subsequent transfer, with the substitution for references in subsection (2) to the purchaser of references to the subsequent transferor" [see CDPA s.56(4)].

Section 56 addresses the situation where the copyright owner of a work in electronic form sells a copy to a purchaser who then transfers the copy to another person (the transferee). At first sight, this section would seem to be of great import for the Internet, particularly if software is made available without any express terms over the Internet; "freeware", for example. However, this rather convoluted section rarely applies where there is a full written licence because of the application of the conditions set out above. In any case, it is still the main practice for software to be distributed with a total embargo upon transfer to third parties and therefore the working of this section is totally avoided.

1.11 Copyright assignment and licensing

Copyright is transmissible by assignment. An assignment can be of the whole of the protected work, or of certain rights relating to that work, for example translation. CDPA s.902(2) states that:

"An assignment or other transmission of copyright may be partial, that is, limited so as to apply –

    (a) to one or more, but not all, of the things the copyright owner has the exclusive right to do;

    (b) to part, but not the whole, of the period for which the copyright is to subsist."

However, CDPA s.90(3) states that:

"An assignment of copyright is not effective unless it is in writing signed by or on behalf of the assignor."

One of the ways in which copyrighted information is protected is through licensing. A licence may be explicit, either in the form of a written agreement, orally or in an electronic message; or implied through the actions of the owner. An explicit licence usually details the uses that may be made of any copyrighted information, particularly relating to the making of further copies. A licence may be implied, for example in the case of putting information on the Web without stating any conditions.

A copyright owner may also grant an "exclusive licence" defined in CDPA s.92(1) as:

"A licence in writing signed by or on behalf of the copyright owner authorising the licensee to the exclusion of all other persons, including the person granting the licence to exercise a right which would otherwise be exercisable exclusively by the copyright owner."

1.12 How to obtain protection	Copyright protection under UK law is automatic. There are no notice requirements or other formalities.
1.12.1 UK statutory presumptions	Although notice of copyright is not a condition of protection under UK law, the CDPA establishes a number of beneficial statutory presumptions if certain notices in the form prescribed by the statute appear on copies of a work.
	For example, where copies of a program are issued to the public in electronic form bearing the name of a purported copyright owner, then that person will be presumed to be the copyright owner.
1.12.2 US/Universal Copyright Convention Notice	Whenever computer software is to be marketed in the US, or is at risk of infringement in the US, it is beneficial to place the US copyright notice on the software product. The US-style notice also satisfies the notice requirements of the Universal Copyright Convention (UCC), thereby obviating the need to comply with any further formalities of protection in other UCC-member countries.
1.12.3 Trade secrets notice	The US/UCC notice can also be combined with a trade secrets notice, so long as care is taken to avoid creating an inference that the software has been "published" for trade secrets purposes. An example of a combined US/UCC/trade secrets notice is as follows:
	"This program is confidential and protected as an unpublished work under the copyright laws of all countries which are signatory to the Berne Convention and the Universal Copyright Convention. Copyright © XYZ Company 1991. All rights reserved".
	This notice should appear on the packaging and at sign-on when the program is in use.
1.13 Infringement issues	CDPA s.16(2) states that:
	"Copyright in a work is infringed by a person who without the licence of the copyright owner does, or authorises another to do, any of the acts restricted by the copyright".
1.13.1 Substantial similarity	References to doing an act restricted by copyright mean doing something in relation to the work as a whole or any substantial part of it, and either directly or indirectly [see CDPA s.16(3)]. In the case of *Total Information Processing Systems Limited v Damon Limited*, 22 IPR 71, 78

(Baker J) the judge stated that the specification of a program cannot be regarded as a "substantial part" of a program within the meaning of CDPA s.16(3).

Assessing whether one work is substantially similar to another work is particularly difficult where the work is a computer program. The similarities may not be as instantly obvious as they would be in the case of a written document and the courts have also taken into account two other aspects - structural similarity and look and feel - which are discussed in the next two subsections.

1.13.2 Structural similarity

In trying to establish whether one computer program is substantially similar to another the courts have been forced to look inside the program at its internal structure to consider whether there is any structural similarity.

In *Ibcos Computers Limited v Barclays Mercantile Highland Finance Limited* (Jacob J, CH 1989 I No 2198, 24 February 1994) at 20 it was concluded that the consequence of regarding [*the plaintiff's program*] as a whole as a copyright work is that it is right to consider what [*the plaintiff's expert*] called the program structure as a whole and the design features as a whole as part of the work, in addition to the literal bits of code and the program structure within an individual program.

In *MS Associates Limited v Power*, [1988] FSR 242, 248 (Falconer J) it was held that the plaintiffs had established an "arguable case" of copyright infringement under the Copyright (Computer Software) Amendment Act 1985 based on a small proportion of "line identities", together with "similarities in structure" and errors appearing in both programs, although denying interlocutory relief on the balance of convenience.

In *Computer Aided Systems (UK) Limited v Bolwell* (Hoffman J, ChD, 23 August 1989), citing *Whelan Associates Inc v Jaslow Dental Laboratory Inc* [1987] FSR 1, the submission was accepted that "the overall structure of the computer program is a form of literary expression in which copyright can subsist", but denying interlocutory relief in the absence of sufficient evidence on what aspects of the plaintiff's program structure had been copied.

1.13.3 Look and feel

Another aspect that the courts have considered, in trying to establish whether one computer program is substantially similar to another, is its external interface. This is termed the "look and feel" of a program; that is, the user interfaces. The courts have made a judgment upon the substantial similarity of these elements of the program and whether it is sufficient to sustain a case for copyright infringement.

In *MS Associates Limited v Power* [1988] 242, 248, Falconer J cited "striking line similarity in the list of functions at the beginning of the defendants' program and its order" as evidence of an arguable case of copyright infringement, although denying interlocutory relief on the balance of convenience.

1.13.4 Infringement law

It is helpful to consider and understand this subject by referring to the results of two cases; Richardson and Ibcos.

The Richardson case

The first report full trial decision on a software infringement case was *John Richardson Computers Limited v Flanders & Chemtec Limited* [1993] FSR 497 (Ch.D). Ferris J held that the plaintiff's computer program for labelling and stock control of prescription drugs dispensed by pharmacists was infringed by the defendant's program, despite the absence of any finding that the plaintiff's code had been copied and despite the fact that similarities which resulted from copying were found in only three features out of a total of 17 features in the two programs.

The Ibcos Case

The second and most recently reported full trial decision on a software infringement case was *Ibcos Computers Limited v Barclays Mercantile Highland Finance Limited* (Jacob J, CH 1989 I No 2198, 24 February 1994). Jacob J held that the defendants had infringed the plaintiff's computer programs for an agricultural dealer system by copying in substantial part both the overall program structure and the individual programs comprising that package, although he found that similarities between certain "design features" did not infringe as those features did not themselves constitute a copyright compilation and, even if they did, the mere taking of those features would not be an infringement as it amounted to the taking of "a mere general idea or scheme." There was no finding that the defendants copied the "look and feel" of the plaintiff's programs; indeed, it was noted that the programs presented themselves very differently to customers.

1.14   Civil remedies

1.14.1 Damages

CDPA s.96(2) states:

"In an action for infringement of copyright all such relief by way of damages, injunctions, accounts or otherwise is available to the plaintiff as is available in respect of the infringement of any other property right".

In the English courts, damages are tried separately from the issue of infringement of IPR. This can appear unfair to the plaintiff who has just won an infringement action and who has to go through the whole process of yet another trial before receiving any compensation. A further disincentive for the plaintiff is that the calculation of the eventual award is difficult to predict due to the dearth of authorities in this area.

The overriding principle in assessing what damages are to be awarded in IPR cases can be easily stated; courts attempt to place IPR owners in the position, as far as money can do it, that they would have been in if their IPR had not been infringed. Over the years, the courts have agreed that IPR owners are entitled, not only to profits on their lost sales and losses caused by price-cutting to compete with the defendant, but also a reasonable royalty rate on sales of the infringing item which the IPR owner would not have made. This last head of damage goes further than the original principle would suggest but it is there that any generosity towards the plaintiff ends.

Other heads of damage, although arguably foreseeable, were considered by courts too remote until the recent case of *Gerber Garment Technology Ltd v Lectra System and another* ([1995] RPC 383) changed the presumption that courts would not award what are called parasitic and springboard damages.

In *Gerber*, Jacob J took a commercial view. He held that lost sales of products associated with the product were recoverable so long as they were caused by the infringement. The judge included associated products, spare parts and servicing in his definition of parasitic damages.

Jacob J also awarded springboard damages; damages to compensate the plaintiff for the fact that the defendant, by infringing the plaintiff's IPR, had gained greater market share than he would have done had he had to develop his own products from scratch. These springboard damages were related to the increased sales by the defendant.

1.14.2 Delivery-up

The copyright owner of a work may apply to a court, for an order that the person who possesses the infringing copy be made to return the work.

This remedy is subject to certain time limits and other conditions specified in CDPA s.99(2)-(4).

1.14.3 Right to seize infringing copies

There is a right to seize infringing copies found to be for sale. This remedy is also subject to a number of conditions specified in CDPA s.100(2)-(6).

1.14.4 Prohibited importation

Under CDPA s.111, the owner of copyright in published literary works (including computer programs) may under certain circumstances give notice to HM Customs and Excise requesting them to treat infringing copies of the work as prohibited goods, in which case importation of such copies is prohibited for up to 5 years and they are liable to forfeiture.

1.14.5 Anton Piller relief

In *Anton Piller KG v Manufacturing Processes Limited*, [1976] Ch 55, the Court of Appeal approved the grant of an *ex parte* order authorising the plaintiff's solicitor to enter the defendant's premises and remove evidence of copyright infringement, provided that it could be shown that (1) the plaintiff has a strong *prima facie* case, (2) the actual or potential damage to the plaintiff's interest is very serious, (3) the defendant is likely to have infringing copies of the work in his possession and (4) there is a real possibility that if the defendant is forewarned he might destroy the evidence.

1.15 Criminal offences

The CDPA establishes a number of criminal offences in CDPA ss.107-110 for making or dealing with infringing copies.

1.16	Crown and Parliamentary copyright	Crown and Parliamentary copyright are special cases of copyright protection, and various of the "normal" copyright provisions are subtly different in relation to Crown and Parliamentary copyright. The specific provisions in the CDPA are set out in s.163-164 (Crown copyright) and s.165-167 (Parliamentary copyright).

1.16.1 Crown copyright

Where a work is made by Her Majesty or by officers or servants of the Crown in the course of their duties then copyright in such work is referred to as Crown copyright. Crown copyright in a literary, dramatic, musical or artistic work subsists:

- in unpublished works, until the end of 125 years from the end of the calendar year in which the work was made

- if the work is published commercially during 75 years from when it was made, for 50 years from the end of the calendar year in which it was published.

Apart from the above, the normal copyright provisions apply; unless the work enjoys Parliamentary copyright protection.

1.16.2 Parliamentary copyright

Copyright in a work is referred to as Parliamentary copyright if a work is made by or under the direct control of the House of Commons or the House of Lords. The House controlling the production of work is its owner.

Parliamentary copyright in a literary, dramatic, musical or artistic work subsists until the end of 50 years from the end of the calendar year in which the work was made.

A work commissioned by Parliament does not become Parliamentary copyright. It is interesting to note the fact that while Acts, after Royal assent, are Crown copyright the text of Bills is Parliamentary copyright.

1.16.3 The role of HMSO

Anyone requiring permission to reproduce Crown copyright material should approach HMSO's Copyright Unit. HMSO administers all Crown copyrights and other copyrights belonging to the Crown. For those Parliamentary items published by HMSO, the Copyright Unit administers Parliamentary copyright on terms and conditions similar to those for Crown copyright.

DEO (PM) (95) 4 dated 12 June 1995[2] sets out the role of HMSO in relation to Crown and Parliamentary copyright. The DEO letter states:

"15. All Crown and Parliamentary copyrights are reserved and will be exercised in appropriate cases. Reproduction will not be allowed in cases of unfair or misleading selection, in connection with advertising or endorsement, nor in any circumstances which are potentially libellous or slanderous of individuals, companies or organisations".

The Copyright Unit has issued general guidelines in the form of its "Dear Publisher" and "Dear Librarian" letters which include general advice to publishers and advise libraries particularly with respect to photocopying arrangements. Copies of both letters are available from the Copyright Unit upon request.

HMSO also licenses other organisations to use and reproduce government material. Most of this licensing activity involves private sector companies wishing to commercially exploit government information in return for payment of royalties.

1.16.4 Moral rights

The right to be identified as author and the right to object to derogatory treatment of a work does not apply to works in which Crown copyright subsists, nor those works in which Parliamentary copyright subsists.

1.16.5 Ownership

In the context of copyright ownership, the right of the Crown relates to its officers and servants. This right differs from the ordinary right of an employer to the copyright in its employees' works in a number of ways:

- the special nature of Crown employment means that the right must apply to office holders as well as employees

- a work qualifies for copyright protection even though the author is not a qualifying person under CDPA s.154

---

[2] A note containing HMSO guidance on the practice to be followed by Departments, Agencies and relevant NDPBs with regard to Crown and Parliamentary copyright.

- Crown copyright attributes are left unaffected by assignment; a work which is originally Crown copyright retains the associated conditions even if ownership is transferred. The reverse is also true of copyrights acquired by the Crown.
  The most significant factor, given the effects of SI1995/3297, is probably the difference in duration between Crown and other copyrights

- where there is a work of joint authorship, one of the authors of which is a Crown servant and one is not, the Crown becomes joint owner but the rights of the other author are unaffected.

1.17 Additional forms of protection

1.17.1 Patent

The Patents Act 1977 s.1(1) states:

"A patent may be granted for an invention in respect of which the following conditions are satisfied, that is to say –

    (a) the invention is new

    (b) it involves an inventive step

    (c) it is capable of industrial application

    (d) the grant of a patent for it is not excluded by subsections (2) and (3) below..."

During the early years of the development of computer technology it was hoped that the law of patent would be the area to afford protection for computer software[3]. During the 1960s software-related inventions became the targets for possible protection by the patent system. But by the early 1970s it had been decided that patent was not the way forward.

Paragraph 487 of the 1970 Report of the Bank's Committee stated that:

"A computer program .... should not be patentable ...".

---

[3] McFarlane. G. "A Practical Introduction to Copyright" 2nd ed 1989, p115

It was difficult to obtain a patent on program-related inventions[4]. Yet the question concerning patents justifies its inclusion in the discussion as it is becoming increasingly an area of interest once more.

In relation to the Internet, patents are becoming increasingly important for protecting non-copyrightable ideas. In 1995 a heavily publicised attempt was made by Compton New Media to patent the search mechanism for a CD-ROM. This would (if the initially granted patent had not been later dismissed) have given Compton New Media a massive monopoly right over a crucial area of new technology.

A requirement for the grant of a new patent is that the invention should differ from "prior art".

When a patent is applied for it should be claimed under a well defined area, so that the patent examiner may discover whether the invention is truly novel and also to ensure that the monopoly is not granted too broadly. In relation to computer programs there is a problem in the application for a patent:

"One of these problems is the incomplete stock of prior art available to patent examiners in evaluating patent applications for processes involving computers, especially those involving software and algorithms ..."[5]

Although there are problems with granting patent protection to computer software there would be advantages if it were granted for such innovations. The essence of patent protection is that it protects the innovation as a whole and so may afford protection to items which would not be protected under copyright. In other words, the processes involved within the innovation which would constitute an unprotectable idea within the copyright system.

---

[4] Bender. D. "Software copyright "look and feel" issues"
Software Protection - vol viii no. 6 Nov 1989, p.8

[5] US Congress, Office of Technology Assessment, Computer Software and Intellectual Property – Background Paper, OTA-BP-CIT-61. (Washington, DC: US Govt. Printing Office, March 1990) p.346

In the patent system a computer program is specifically excluded from protection by statute but if the program were incorporated in a fixed form into the hardware then the program might be protected as the novel part of the invention. This potential loophole is not so wide as may first appear. If a program were incorporated into the hardware of the computer then it would cease to be a piece of software and would instead be much nearer to the classical subject matter for patent protection.

Patent has been heralded by some as the new form of software protection for the 1990s. There are however, arguments for and against this proposal and these have been enunciated above. Additionally the problems of fitting patents to computer technology restricts the patenting of computer software. It has also been argued that patent protection would restrict software development[6].

There do exist other alternatives to copyright rather than the perhaps obvious alternative of patent protection.

1.17.2 Contract law

Contract law is an invaluable method of protection but only between the parties to that contract. A contract is an agreement between two or more parties who have an intention to create legal obligations between themselves. A third party to a contract is unprotected from any breach of the contract terms. The period of protection provided by contract law varies and depends upon the period stipulated in the terms of the contract. A contract is breached if one of its parties acts contrary to one of the contractual terms.

Contract law does not provide an adequate means to prevent the copying or unauthorised use of a copyright work by a third person. "Shrink-wrap" licensing has attempted to deal with this problem by trying to establish a contract between a software producer (rather than the retailer) and the customer. The technique means that the buyer is taken to have accepted all the terms of the licence when the seal of the program package is broken.

---

[6] Charles. D. "Rights and Wrongs of Software" New Scientist 29 Sept. 1990, p48

However, it is not:

"... entirely clear whether the practice ... constitutes a valid licence in all circumstances and in all jurisdictions."[7]

Contract law can be of use between one software house and another; for example, in making a formal agreement to collaborate upon the development of new product. Because of the expense involved, using this sort of software generally involves a written contract.

Contract law is not, however, a useful remedy to stop infringement of copyright. This as previously stated is due to the reliance of contract law upon Privity of Contract - in other words contract is only effective against parties to a contract who have Privity of Contract. These problems may be redressed by the ability of the law of confidentiality to afford relief against non-contractual parties, particularly where negotiations lead to no concluded contract.

1.17.3 Trade secrecy and confidentiality

Whereas copyright and indeed patent are almost entirely based upon statute, the law relating to trade secrets is generally lacking in any settled contractual base. Despite this uncertainty, the scope of protection offered by both trade secrets and confidentiality is potentially very wide. The width is because there are no restrictions in terms of "ideas" or "expressions" as to what may be termed confidential information. An idea may be protected as easily as the expression of that idea[8].
An action for a breach of confidence (trade secrets) protects against the disclosure of confidential information in or about an asset by someone under an obligation of confidence[9] (see also 9.5).

Trade secret law might also be a desired alternative for reasons including speed and cost[10]. In other words the

---

[7] Source Materials "Proposal for a Council Directive on the Legal Protection of computer programs" Software Protection, April/May 1989, p.15

[8] Millard. C.J. "Legal Protection of Computer Programs." (1985), p.119

[9] Niblett. B. "Legal protection of computer programs." (1980), p.114

[10] Tapper. C. "Computer law" 4th ed 1989, p.83

protection afforded is immediate, requiring no formalities and may continue indefinitely. Trade secrets help in the period before any protection has been granted; the period before the idea has been incorporated into the hard-wiring of a computer, or before it has been expressed in any tangible form.

Specifically, a common situation in questions of confidentiality may occur whereby new ideas are disclosed to potential customers prior to the formation of any contract. In such situations contract law would be of no use. The law of confidence provides a remedy where a customer has gained a degree of unjust enrichment (in terms of free information) at the expense of the developer. Confidence law also helps to protect the developer against unlawful actions by those who are not direct parties to any contract which may be in existence[11].

There are however, problems with reliance on trade secrecy and confidentiality to protect valuable information.

"While the misappropriation of specific secrets,... may be actionable, an employer cannot prevent a former employee from using his or her general skills and experience when moving on to a new job"[12].

The software industry, for example, features a high turnover in manpower and so there will be similarities in style between a program written by a programmer whilst with one firm and with a later program written when with another. It is clear that other forms of protection are more useful in detailing with this quirk of the software industry.

---

[11] Tapper. C. "Computer law" 4th ed 1989, p.99

[12] Millard. C.J. "Legal Protection of Computer Programs." (1985), p.120

1.18 Implications for the Internet

1.18.1 What is copyrightable?

Information on screen is copyrightable (see 1.6.1) provided the various criteria (such as originality) are present in order for copyright to apply. On the Internet, information is accessed and, in order to see that information, it must be copied (temporarily at least) onto a computer. Therefore if information is put onto the Internet, for example on a Web page or bulletin board, without the permission of the rights holder, the copyright in that information may be infringed.

On-line distribution must include initial storage (however briefly) and possibly subsequent storage by the recipient, both potential infringements which, whilst not expressly set out in the CDPA, may be inferred from reading section 16 of the Act.

Under s.18 of the CDPA, the distribution right affords the copyright owner the exclusive right to issue copies of a work to the public. It is debatable whether on-line transmission actually constitutes such a distribution when in reality it is not a "copy" of the work which is being issued to the public. Digital transmission means that the actual work, and not a copy of it, is transmitted. Therefore, there is currently wide debate as to whether a new "transmission right" should be introduced to close this gap.

It is thought likely that the European Union will further examine the need for a digital transmission right as a result of the 1995 Green Paper consultations. Existing models such as the provisions of the EU Rental Right Directive and the EU Software Directive are unlikely to prove conceptually sufficient to deal with digital transmission. The conceptual difficulty lies (with respect to rental or lending) with the fact that traditional forms of lending or rental envisage the parting with possession of an article. The new form of transmission means that the lender still retains the "original" work whilst allowing the "lendee" to also enjoy the work.

At the same time there are other schools of thought which hold that there is no logical or legal difference made to what constitutes a "copy" merely because the activity concerned is digital transmission of a work. Copying is a process that produces two or more sets of data where previously there was only one. Just because digital transmission is such a technically efficient process and that the result is not just corresponding or similar but identical in every respect, this makes it no less a process of copying. It would not be a copy only if the original data was moved in the course of transmission from where it originally was to a remote place, leaving it as still the only existing version.

Therefore putting copyright works on the Internet may infringe the rights of the copyright owner, as may the person who actually transmits that work (see 1.6.2).

1.18.2 Duration of copyright

Section 1.4 above sets out the general rules concerning the term of copyright protection. The important point to note here is that electronic products and services containing a variety of content (for example a multimedia product containing film, text and sound), can blur the distinctions between the different sources of copyright. It may well be that the copyright in the individual elements of the product expire at different times. This is an example of the potential for anomalies caused by the impact of the information society on copyright law.

1.18.3 Copyright licence

Placing information on a Web site, without incorporating any licence conditions, may be regarded as giving implied licence to any person downloading the information, or even calling information up on the screen. In the latter case the person has copied the information, albeit briefly, onto the computer's RAM and there is an argument that this gives implied licence to have further rights to copy. However, the implied licence argument, in relation to information put on the Internet, is not universally accepted.

1.18.4 Works made in employment

As section 1.3 has already stated, there are general presumptions as to first ownership of copyright. It is important therefore to ensure that appropriate mechanisms are in place to deal with the ownership of the rights in content. For example, organisations may wish to put information onto a Web site and the information may have come from a number of sources; external developers and consultants, internal employees, etc. Organisations therefore need to secure releases of rights from any third parties, and be sure that any employees really are creating the content during the course of their employment.

1.18.5 Devices to circumvent copy protection	CDPA s.296 (see 1.6.3) sets out provisions which make a person selling a device to enable another to circumvent copy protection, liable for copyright infringement. These provisions clearly catch devices sold to deliberately decrypt encrypted messages or information.
1.18.6 Transmission of works	Works which have been put into electronic form are easy to transmit. The Internet enables the transmission of works in electronic form more than ever before. The normal rules apply to such transmissions (see 1.6.2), subject to the results of the European Commission's Green Paper.
	The Commission have particularly asked for views on the creation of a new "transmission right". This is due to the blurring of the traditional concepts of what constitutes an original work and what constitutes a copy of that work and the difficulty of assessing this question in practice (see 1.18.1).
1.18.7 Liability of operators	The question as to whether operators of bulletin boards or Internet service providers can be liable for copyright infringement because of the actions of their subscribers, even if they have no knowledge of what their subscribers have been doing, has not yet been tested in the English courts, but it would seem that they may be liable. If any material passes through an operator's computer it will almost certainly be deemed to have been "copied" in the sense in which the word is used in copyright law, therefore making the operator subject to the principle that copying is a "primary infringement".
1.18.8 Crown copyright	The considerations outlined elsewhere in this section apply equally to Crown copyright material. Ease of manipulation of electronic data should be the main concern for organisations wishing to put information onto the Internet.
1.18.9 Conversion of program code	Given the rules set out at paragraph 1.6.1. concerning adaptation and program conversion, it is necessary to be able to distinguish between a translation from paper text to Hypertext Mark-up Language (HTML), not an

infringement in principle, and a translation from one program to another which constitutes an infringement if unauthorised by the rights holder.

1.19   The challenge of new technology

This chapter has raised a number of issues where the current law of copyright is stretched beyond its originally intended application. For example, the blurring of the distinctions between various copyright works when those works are reduced to electronic form and bundled together as a multimedia product.

The fundamental issue was identified in the Bangemann Report[13].

"...the ease with which digitised information can be transmitted, manipulated and adapted requires solutions protecting content providers. But at the same time, flexibility and efficiency in obtaining authorisation for the exploitation of works will be a pre-requisite for a dynamic multimedia industry".

Perhaps the most pressing issue is that of enforcement.

Electronic reproduction threatens the perception of control of rights owners. There is uncertainty over how to regulate and charge for information made available over networks. Much effort is being invested to install technical means of detecting infringing copies of electronic products and services, but there is a perception that such efforts are being outstripped by technical advances.

Standard means of distribution of copyright protected works are being circumvented by transmission routes: distribution of academic journals is threatened by online distribution of electronically stored journals; rental and sale of video cassettes is threatened by the provision of on-line interactive TV. It is unclear what role contract and licensing can play in seeking to control the activities of third parties who acquire unauthorised electronic copies of works.

---

[13] A European Council requested Report looking at Europe and the global information society: 26 May 1994

Mechanisms have been proposed, including digital tagging of data to enable the movement of valuable data to be traced. For example, the United States Information Infrastructure Task Force produced a White Paper in 1995 entitled "Intellectual Property and the National Information Infrastructure". The paper is the report of the working group on Intellectual Property Rights, chaired by Bruce Lehman; the Assistant Secretary of Commerce and Commissioner of Patents and Trademarks.

The White Paper aims to clarify and update the existing US copyright law. One of its aims is to ensure that copyright management information is attached to electronic copies of a work to ensure that rights holders and publishers can track every use made of electronic copies. In conjunction with digital tagging it is proposed that works in electronic form are protected in some way; for example encrypted, and that attempts to circumvent copyright protection be made illegal.

2   Defamation

2.1 Introduction	Defamation is the capability to cause damage through idle or intended words. The Internet, as a microcosm of the information society at large, presents a whole new and global way in which to such damage. Individuals who issue defamatory statements over the Internet are likely to be just as liable as if they uttered them on a public platform or published them in a newspaper.

It is an example of the all-reaching nature of the Internet that suddenly IT lawyers need to become conversant with the civil law principles of defamation.

This chapter examines the basic tenets of the applicable law and applies those principles to the new media; the new technology. This chapter looks at the proposed developments in this area of law, sparked mainly by technological developments.

Liability of the various "players" (for example Internet service providers, bulletin board operators and those who post messages to such boards), are determined by the role that each player assumes. As is discussed below, bulletin board operators may have a different legal responsibility depending upon whether or not they are exercising editorial control. The position of Internet service providers should be made clearer if the proposed amendments to the Defamation Act 1952 become law.

2.2   Sources of the law

The law in England governing defamation is complex. It has its roots in common law, and, as such, has often struggled to come to grips with the issues raised by changing circumstances and advances in technology. Sporadic attempts have been made by legislation to solve some of the problems of the common law of defamation, the most important of which is the Defamation Act 1952, itself a frequent object of criticism.

Fresh legislation is proposed to update and improve the 1952 Act, but these proposals, too, are not generally considered comprehensive enough. This chapter returns to these proposals later.

2.3	**Basic principles**	The essence of the law of defamation is the publication of matter conveying a defamatory imputation. Only the subject of the defamatory publication acquires a right of action and may sue; for example, a person whose name has been slurred.
2.3.1	Test of defamation	There is no single satisfactory test of whether or not a statement is defamatory. Common formulations include whether the:

- statement tends to expose the subject to hatred, contempt or ridicule

- statement tends to cause others to shun and avoid the subject

- statement tends to lower the subject in the estimation of others

- statement is to the subject's discredit.

The statement must contain a statement of fact or expression of opinion or imputation conveyed by it having the defamatory effect. It is not defamatory to identify the person as the subject of a previous defamatory statement, even though this could lower that person in the estimation of others.

The fact that the statement in question does not actually cause the subject to be regarded with hatred, or cause others to shun or lower their opinion of the subject is irrelevant, it must merely tend to do so. In judging this, the standard of opinion is that of right-thinking members of society as a whole; that is, what would right-thinking members of society think of the defamed were they to hear the defamatory statement? If it is thought that right-thinking members of society would think worse of the defamed, then the statement made becomes defamatory.

2.3.2	Requirement of publication	The defamatory statement must be published to a third party, and not simply to the subject of the statement; therefore if A defames B, B and C must know about it.

The words must convey the defamatory meaning. If a defamatory statement is spoken in a foreign language not understood by those present, there is no publication. (Clearly, this is not the ordinary use of the word "publication"; it is closer to "communication" where at least one person other than the defamed person hears or reads the remarks.) Equally, the words must convey a meaning defamatory to the subject.

As a general rule, where a letter is addressed to a particular person, the writer is not responsible except for publication to that person. However, where the writer knows that the letter is normally opened and read by someone other than the addressee (for example, a secretary who always opens the post), this can constitute publication to a third person.

In the cases of postcards and telegrams, there is a presumption of publication, since these items are by their nature likely to be read by third parties. However, this presumption does not apply where defamatory matter is sent in an unfastened envelope.

The author is also liable where the defamatory material is published unintentionally, unless it can be shown that the publication was not due to any lack of care on the author's part.

| 2.3.3 | Distinction between libel and slander |

The distinction between libel and slander can be a difficult one to draw. Broadly speaking, if the publication is made in permanent form or is broadcast, the matter is a libel; if in an impermanent form, it is a slander.

In common law, any publication of defamatory matter in permanent form is a libel. "Permanent" does not mean long-lasting; for example, burning a man in effigy and cinema films have been held to be libellous.

With the advances in technology, the distinction between libel and slander, resting as it does in common law on the question of whether or not the defamatory material was in a permanent form, has become increasingly difficult to apply.

The Defamation Act 1952 provides that the broadcasting of words for general reception by means of wireless telegraphy shall be treated as a libel. Broadcasting other than for the purposes of general reception is still covered by the common law rules.

The distinction is important because, in the case of libel, the subject is presumed to have suffered loss and need not specifically prove the loss. By contrast, if the plaintiff has been slandered, he or she must show that some form of special, actual damage capable of being estimated in money (for example, the loss of a job or business, or even the withdrawing of a dinner invitation) has been suffered. Certain slanders, however, are presumed to cause special damage. These are the imputation of:

- a crime

- contagious disease

- un-chastity in women (Slander of Women Act 1891)

- lack of fitness for one's trade or business. (s.2 1952 Act).

2.3.4 Defences

There are two classes of defence to an action for libel or slander. The first is arguing that the elements of libel or slander are not present. In other words that the words were never published; the words were not directed at the plaintiff; the words did not mean what it is alleged they meant; or in the case of slander, that there has been no special damage, or such damage was too remote.

The second involves accepting that a statement has been made which is prima facie defamatory, but argues that for one or more of a variety of reasons, no liability should arise. The most common of these defences are as follows:

- justification

- privilege (absolute and qualified)

- fair comment

- unintentional defamation.

Each of these defences are examined in more detail below.

Justification

With one exception, it is an absolute defence to prove the truth of the imputation complained of.

In the case of a general charge, one must show that the charge was substantially true; in the case of a specific charge, one must prove the truth of the charge (though small discrepancies of time and place are permissible).

Under Section 5 of the Defamation Act 1952, the defendant can rely on partial justification in mitigation of damages. Therefore, if a statement contains two or more charges against the plaintiff, a defence of justification cannot fail by reason only that the truth of every charge is not proved if the words not proved to be true do not materially injure the plaintiff's reputation having regard to the truth of the remaining charges.

The one exception to the rule that justification is an absolute defence, regardless of motive, is in respect of "spent" offences under the Rehabilitation of Offenders Act 1974. Under that Act, certain convictions are effectively wiped from the slate after a number of years. This raises obvious problems in the area of defamation. The Act confirms that the defendant may generally rely on evidence of spent convictions in support of a defence of justification, but provides an exception where the publication was proven to have been with malice.

Absolute privilege

A defence of absolute privilege attaches to defamatory statements:

- made in the course of judicial proceedings (including court hearings, pleadings and other documents brought into existence for the proceedings and proofs of evidence given in court which the witness has given to a solicitor beforehand)

- made in the course of quasi-judicial proceedings; for example, commissions and enquiries

- contained in documents made in judicial or quasi-judicial proceedings

- made by one Officer of State to another in the course of official duty

- made in the course of parliamentary proceedings

- contained in reports published by order of either House of Parliament

- contained in reports of the Parliamentary Commissioner and in communications with a member of the House of Commons for the purposes of the Parliamentary Commissioner Act 1967

- made in fair and accurate reports in a newspaper of proceedings publicly heard before a court exercising judicial authority within the UK (this has been extended by Section 9 of the Defamation Act 1952 to fair and accurate reports of such proceedings for broadcast purposes and to such broadcasts).

Absolute privilege is an absolute defence, regardless of the motive behind publication or the truth or otherwise of the statement.

Qualified privilege

There are occasions where, on grounds of public policy, a person may make statements about another which are defamatory, but which incur no liability.

On these occasions, if a person states what he or she believes to be the truth about another, he or she is protected in doing so, provided he or she makes the statement honestly and without any indirect or improper motive (see the sub-section on malice below).

The occasions where the defence of qualified privilege may arise, being based on public policy, are not set. Generally, the most important are where statements are made:

- in the discharge of a public or private duty (whether legal or moral); or

- on a subject matter in which the defendant has a legitimate interest.

For the defence to arise, the statement must be made to someone who has a legitimate interest in the matter (and not merely where that person is casually interested in it).

Fair comment

It is a defence that the words complained of are fair comment on a matter of public interest. For the defence to apply, the comment must be made honestly and without malice.

The defendant must show the following:

- that the publication was a comment, not a statement of fact

- that there was a basis of fact for the comment complained of

- that it was a comment on a matter of public interest.

It is not acceptable to make a defamatory misstatement, no matter how honest, nor how reasonably the maker believed in the truth of the misstatement.

Malice

The defences of qualified privilege and fair comment, depending as they do to a large extent on the state of mind of the defendant, will not be successful where the plaintiff can show that the defendant was not acting honestly, but was actuated by actual malice. Malice need not be personal spite or ill-will, merely some indirect or dishonest motive.

Some general rules were laid down in the case of *Horrocks v Lowe*[14], which was a case involving qualified privilege, but should also be treated as applying to fair comment. The most important of these rules are:

- defendants are entitled to be protected unless some dominant and improper motive on their part is proved

- knowledge that a statement injures the plaintiff does not destroy the privilege if the defendant was using the occasion for its proper purpose

- generally, though not always, lack of belief in, or indifference to, the truth of the statement is conclusive evidence of malice

- generally, though not always, belief in the truth of the statement protects defendants, unless they are shown to have misused the occasion.

---

[14] [1975] A.C. 135

Unintentional defamation Section 4 of the Defamation Act 1952 provides a defence in certain cases of unintentional defamation. For this to succeed, the defendant must show:

(1) an offer of amends has been made; and

(2) either:

(a) the offer has been accepted and performed; or

(b) (if the offer has not been accepted) that the matter was published innocently, the offer was made at the correct time and has not been withdrawn and (where the publisher is not the author) the words were written by the author without malice.

Section 4(5) of the 1952 Act provides that the matter is to be treated as published innocently if the publisher took all reasonable care in publishing it, and one of the following two conditions is satisfied:

(1) the publisher did not intend to publish the words about or concerning the plaintiff, and did not know of the circumstances by virtue of which they might be understood to refer to the plaintiff; or

(2) the words were not defamatory on the face of them, and the publisher did not know of circumstances by virtue of which they might be understood to be defamatory of the person.

The offer of amends has to be made as soon as practicable, and consists of:

(1) an offer to publish or join in the publication of a suitable correction and suitable apology to the aggrieved party; and

(2) an offer to take reasonable steps to notify people in receipt of copies that the words are alleged to be defamatory of the aggrieved party.

2.4 Liability of others Generally speaking, liability for publication arises from participation or authorisation in the defamation. Therefore, in the case of a newspaper, all those who have taken part in publishing, or have submitted material in it are *prima facie* liable. The analysis is not so simple when the medium of publication becomes the Internet (see 2.5 below).

Many editors, printers, publishers and vendors have been held liable for defamation. So too have proprietors of newspapers, since editors are their servants, and it is within the scope of their employment to send to printers whatever matter they think should be published. This illustrates the effect of the general principle of vicarious liability, whereby employers may be held liable for torts committed by their employees in the course of their employment. This same principle applies to on-line dissemination of information.

The author of a defamatory statement is liable in the normal course of events. Where the author becomes part of the chain of publication, he or she may also be liable if it is shown that publication was requested or authorised.

2.4.1   Republication

Every republication of a libel is a fresh libel.

Originators are generally not liable for the voluntary and unauthorised repetition and republication. They are liable where:

- they authorised or intended the person to whom they published the words to repeat or republish them to some third person

- repetition or republication was the natural and probable result of the original publication

- they were under a moral duty to repeat or republish the material to a third party.

2.4.2   Distributors

The distinction between publisher and distributor is key to the analysis of liability of the various persons involved in an on-line defamation scenario.

Where a person is not the author, printer or "first or main publisher", but rather is involved in a subordinate role, there is no liability if it can be shown that:

- the person did not know the publication contained the libel complained of; and

- did not know the publication was of a character likely to contain a libel; and

- such lack of knowledge was not due to any negligence on their part.

The mere fact that the person has not read the publication in question right through is not evidence of negligence.

2.5 Implications for the Internet

2.5.1 Libel or slander

As mentioned in 2.3.3 above, the most important result of the difference between a libel and a slander is that plaintiffs have to prove they have suffered special damage in the case of slander, whereas damage is presumed in the case of libel.

It is likely that defamatory material in electronic form will constitute libel rather than slander. Material is stored, even if only temporarily, in electronic form, and material broadcast for general reception (analogous arguably to the on-line scenario), is stated by the Defamation Act 1952 to be libel. While the question has not been specifically addressed in litigation, the few cases on defamation in the context of the Internet, both in the UK and the US, have proceeded on the assumption that the defamatory statements were libellous rather than slanderous. This is obviously an advantage to the plaintiff in any defamation action.

2.5.2 Publication

Clearly, sending an e-mail to a third party as well as to the recipient, constitutes publication by the author, as does placing a message on a bulletin board. The position becomes less certain when one considers the issue of e-mails sent to only one person, and that person is the subject of the defamatory matter. Is this to be taken as a publication to anyone who has the ability to read e-mail traffic? It is unclear whether this should be treated in the same way as postcards (publication presumed) or unsealed envelopes (publication not presumed) and perhaps such analogies are not helpful.

In the absence of case law on this point, however, the safer option would be to encrypt the message. This at least, would show a positive intention not to publish the message and arguably, if a case went to court, could be good evidence of an absence of malice; perhaps enough to preserve the various defences open to the defendant.

A defence of absolute privilege attaches to statements made by one Officer of State (that is, a public servant) to another in the course of official duty. This is of particular use and interest to Officers of State as it means that, for example, their e-mails to other officers are immune to the normal rules of defamation.

2.5.3 Liability of participants

Of wider concern than the liability of authors of defamatory material is that of other participants in networks. Providers of the networks carrying electronic messages, Internet service providers, bulletin board operators and, owners of computers through which e-mails may be routed, all risk having actions brought against them for publishing or republishing defamatory matter.

This is likely to be of concern given the comparative ease with which authors may make themselves anonymous[15], and the fact that subjects of defamatory statements are often tempted to sue parties which they perceive to have "deep pockets", such as the systems operators, or parties which are based within jurisdictions with strict rules on defamation.

Anyone who has participated in, or authorised, publication is *prima facie* liable. However, someone who has taken only a subordinate part in distributing a libel may have a defence (see 2.4.2). There is therefore an important distinction between being regarded as a publisher (and therefore liable) or an innocent distributor (and therefore potentially not liable).

Network and local access providers

In the US, telecommunications operators are protected from defamation lawsuits as "common carriers" if they are doing nothing more than providing the lines over which information is transmitted. No such protection exists in the UK (though legislation is proposed which would rectify this - see 2.7). However, as observed in the Introduction to this Chapter, liability depends upon the activity being undertaken by the telecommunications operator. Setting up discussion forums, for example, introduces potential for liability and removes the "common-carrier" exemption.

[15] There exist sites, known as "anonymous re-mailers", which strip the sender's identity from an e-mail or posting

Nonetheless, it seems likely that telecommunications operators providing a telephone service can argue that they are only distributors of the defamatory material (see 2.4.2).

This defence also appears to be open to local access providers who are merely providing the network structure.

Providers of "passive" hosts and managers of "active" groups

The position is more complex in relation to providers of bulletin boards.

Providers of "passive" hosts tend to argue that they are in the same position as a telecommunications company, since they merely provide the means by which other parties communicate, and are therefore an innocent distributor. However, it is arguable that in the case of material on a bulletin board the operator is in fact a "primary" publisher rather than a distributor, regardless of any fault.

"Active groups" are more interactive and feature participation by the groups' providers. Managers of "active groups" tend to be more involved in the content of the material in their groups, and therefore it is harder for them to argue that they should be treated as passive conduits. Whilst they may seek to argue that they are merely innocent distributors, it is more difficult for them to argue that they are unaware or should not have known that any of the material contained was likely to be defamatory (see 2.6).

Internet publishers

Similar questions arise in the context of the Internet as a whole. Hosts through whose machinery messages pass seem likely to be treated in the same way as network providers (leaving aside the problem of establishing the route taken on the Internet by any particular message). On the other hand, providers of Web pages seem more likely to be held to have intended to publish and to have published the material on the pages. This seems right, as Web publishers have actively put the information onto their pages.

Providers of HTML links to defamatory material

Hypertext Mark-up Language (HTML) provides a service whereby Web pages can inter-link. In HTML certain key words are highlighted. By clicking on these, one can access relevant parts of the text connected to the highlighted words. If those parts of the text contain defamatory material, it is arguable that the HTML link providers have published such defamatory material, and/or that the Web page provider has drawn attention to defamatory material, causing the Web page provider to become liable for the defamation.

2.6   Case law

There is very little UK case law on e-mail and Internet libel, and none reported on the position of parties other than the author of the libel.

Two cases in the US, however, are of interest.

Cubby v CompuServe[16] ("Cubby")

CompuServe, an on-line service provider, ran a variety of special interest forums. An independent third party had contracted to manage, review, edit and control the contents of the forum.

A daily newsletter called Rumorville was published in the forum, which contained an allegedly libellous remark concerning *Cubby*. *Cubby* sued various parties, including CompuServe.

CompuServe argued that it was a distributor, rather than a publisher. The Court agreed with this argument, noting that "CompuServe has no more editorial control over such a publication than does a public library...". There was no evidence that CompuServe knew or had reason to know of the libel, and therefore the claim failed.

Stratton-Oakmont Inc v Prodigy Service Co.[17] ("Prodigy")

By contrast, *Prodigy*, also an on-line service provider, was held to be a publisher of libellous material. The Court focused on the fact that *Prodigy* held itself out as exercising control over the content of its on-line services. Indeed *Prodigy* advertised the fact that it was running a "family" on-line service. The Court drew a distinction between "passive conduits of information" and those who exercise editorial control, which brings with it

---

[16] 776 F.Supp. 135 (S.D.N.Y. 1991)

[17] NY Sup Ct May 24,1995

increased liability. Ironically then, from the defamation point of view, it would seem better to exercise no control than to exercise a degree of control.

While *Cubby* and *Prodigy* have been commented on extensively in the UK, one should not lose sight of the differences between the likely approaches of courts in the US and in the UK.

The US courts took the view that the critical issue was the degree of control exercised. In the UK, the question is whether the defendant is a primary publisher or a distributor. If the defendant is a primary publisher, the issue of control is wholly irrelevant.

Whether or not a party is a distributor is influenced by the degree of control exercised. By definition, a distributor merely distributes, rather than controls the content of what is distributed, however, it is not possible to be so unequivocal with regard to on-line information.

2.7 Proposed amendments to the Defamation Act 1952

2.7.1 Provisions of the Bill

A draft Bill has been published amending the Defamation Act 1952. In the context of defamation and the Internet, the most important provision is contained in Article 1, which provides that in proceedings for defamation it is a defence to show that a person was not primarily responsible for the publication of a statement complained of, did not know, and having taken all reasonable care had no reason to suspect, that a person's acts involved or contributed to the publication of the defamatory statement.

Under the Bill the following shall not be treated as being primarily responsible for the publication of a defamatory statement:

(a) in the case of a defamatory statement published by electronic means, a person involved only:

(i) in processing, making copies of, distributing or selling any electronic medium in or on which the statement is retrieved, copied or distributed, or

(ii) in operating any equipment by means of which the statement is retrieved, copied or distributed

(b) the operator of a communications system by means of which a defamatory statement is transmitted, or made available, by a person for whose acts the operator is not responsible.

In cases to which the provisions of Article 1 do not apply, the courts shall apply them by analogy.

In determining whether there is reason to suspect that a person's actions involved or contributed to the publication, regard shall be given to whether, among other things:

- the nature or circumstances of the publication, or

- the previous conduct or character of those primarily responsible,

were such as to give cause for suspicion that such actions might involve or contribute to the publication of a defamatory statement.

2.7.2 Effect of the proposed amendments	Network providers clearly fall within the scope of those not primarily responsible for publication, as are those people through whose computers messages travel on the Internet.

The position of a bulletin board operator is still not entirely clear, however. Arguably, a provider of an entirely open bulletin board is not primarily responsible. This class, however, does not appear to cover providers of bulletin board services who have any form of control over the content of the bulletin board.

Ironically, the impression may be that it is better not to exercise any control over a bulletin board for fear of being held liable for its contents, but as the above has shown, the issues are more complex and much depends upon the role being assumed.

3　Data protection

3.1　Introduction

The Data Protection Act 1984 (DPA) exists as a result of general concern over the impact of computer technology on personal privacy.

Developments in computing have meant that more and more data can be stored, manipulated and sold. The advent of the Internet heightens such concerns. Data about people can now be transferred very quickly and to almost any destination globally. Further, with so many people involved, electronic litter is scattered around the Internet and an alarming amount of information is there for the taking, sexual orientation, bank balances, political persuasion, etc. It's all there for the world to see and for business to read and sell.

The main effects of the DPA are to confer rights on individuals about whom information is held on computers and to impose obligations upon those who handle that data. The DPA is a set of legal rules forming a subset of the concept of the right of privacy of individuals. Privacy is a concept that is apparent in the national legal systems of several continental European states, however, it is not a familiar concept in English law.

An EU Data Protection Directive has recently been adopted. Its thrust is to stop the unfair use of personal data and the processing of that data. Processing is a very broad term. Processing within a computer, even if it is just extracting electronic information, is still processing for the purposes of the directive and is subject to examination as to whether or not that processing is fair. Similarly, under the terms of the Directive, the display of information on a screen constitutes processing.

It is generally agreed that the Internet increases the risk of unfair use of personal data especially given its global nature. Although the provisions of the EU Directive attempt to harmonise the position between member states, communications over the Internet are not restricted to the member states. The problem is that non-EU countries may have a lower standard of data protection for personal information.

To assist in interpreting the DPA, the Data Protection Registrar has issued a useful set of Guidelines (the current version is the Third Series published November 1994), which clarify and enlarge on the DPA. In addition, there is a small body of case law on the question of data protection.

To put the issue of privacy in context, according to the BBC World Service a woman in the US, who had escaped an abusive husband and started a new life, was tracked down by her ex-partner; apparently through her use of an on-line service with Internet access. The story may be well on its way to becoming an urban myth but it does highlight an often ignored and usually disregarded aspect of the Internet; electronic footprints are left wherever one steps in cyberspace and it does not take an electronic Sherlock Holmes to follow the trail.

3.2 Some relevant definitions from Section 1 of the DPA

Personal Data

Information:

(a) relating to a living individual, provided that individual can be identified from the information, either on its own or in conjunction with other information in the possession of the "data user"; and

(b) which is recorded in a form in which it can be processed by equipment operating automatically.

While the personal data must itself be held in a processable form, other information which enables one to identify the person need not be. Thus, if the personal information is referenced to a number, and the "data user" has a list in manuscript enabling a correction to be made between the name and the number, the personal information is regarded as "personal data".

Data Subject

The individual about whom the personal data is held.

Data User

The person who holds the data.

Holding Data	A person "holds" data if:

    (a) the data is part of a collection processed or meant to be processed by or on behalf of that person; and

    (b) that person controls the content and use of the data; and

    (c) the data is in a form which is, has been or is intended to be processed as in paragraph (a) above.

Computer Bureau	An organisation carries on a business as a computer bureau if it provides other people with services in relation to the data.

An organisation provides services in relation to data if:

    (a) as an agent for other persons it causes data held by them to be processed by equipment operating automatically; or

    (b) it allows other persons to use equipment in their possession to process data held by them.

Processing	"Processing", in relation to data, means amending, augmenting, deleting or re-arranging the data or extracting the information constituting the data. This is an exhaustive list.

"Processing", in relation to personal data, means processing by reference to the data subject.

Disclosing	"Disclosing" in relation to data, includes disclosing information extracted from the data; and where the identification of the individual who is the subject of personal data depends partly on the information constituting the data and partly on other information in the possession of the data user, the data shall not be regarded as disclosed or transferred unless the other information is also disclosed or transferred.

3.3	**Data protection Principles and the Internet**	The DPA is based around eight Data Protection Principles contained in Schedule 1 to the Act. These Principles encapsulate good practice with regard to the protection of personal data by data users. They are designed to comply with the principles as stated in the Council of Europe Convention on Data Protection.

3.3.1 First Principle
first part

"The information to be contained in personal data shall be obtained fairly and lawfully"

In some cases, information is always to be treated as being obtained fairly. These are when the person from whom it is obtained:

- is required or authorised by statute to supply it

- is required to supply it by a convention which imposes an international obligation on the UK.

Any disclosure of information which is so authorised or required must be disregarded when considering whether the information was obtained fairly. In judging whether the information has been fairly obtained, the relevant time is that at which the information was obtained (see *Innovations (Mail Order) Limited v Data Protection Registrar* Data Protection Tribunal DA/92 31/49/1).

The DPA gives a special dispensation from this Principle for personal data held for historical, statistical or research purposes. Information is not to be regarded as obtained unfairly merely because its use for those purposes was not disclosed when the information was obtained. This dispensation applies only if the personal data is not used in such a way that damage or distress is, or is likely to be, caused to any data subject.

Registrar's Guidelines

In considering whether information has been fairly and lawfully obtained, the Registrar takes into account all the circumstances and considers issues such as:

- was it reasonable to expect the person supplying the information to appreciate, without any further explanation by the data user, the identity of the data user and the purposes for which the information would be used or disclosed? If not, why did the data user not explain them to the person?

- if the data user did explain why the information was required and why it might be used or disclosed, was the explanation complete and accurate?

- did the person ask about uses and disclosures of the information and, if so, what reply was made?

- was the person supplying the information under the impression that it would be kept confidential by the data user? If so, was that impression justified by the circumstances and did the data user intend to preserve that confidence? (Is a data user who puts data on-line still preserving that confidence? This depends both upon the extent of that data user's registration and the extent of security arrangements. For further discussion of security issues, see Chapter 9).

- was any unfair pressure used to obtain the information; for example, were any unjustified threats or inducements made or offered?

- was the person improperly led to believe that he or she must supply the information, or that failure to provide it might disadvantage him or her?

- did the data user have any particular knowledge about the person from whom the information was obtained, either because the person was one of a specific group, for example young people, or because the data user had a personal relationship with the person? If the data user had such knowledge, would the explanation given by the data user concerning the collection and intended uses be understood by the ordinary man in the street?

If a data user requests information about individuals either from themselves or from others, the data user should try to make sure that no-one is misled as to who the information is for, why the information is required or why it is to be used or disclosed. The Internet, through use of interactive Web pages, for example, is a prime opportunity for organisations to gather personal data should they so require. The principle of "fair obtaining" should therefore be borne in mind.

The data user must always be fair to the person from whom the information is obtained. Fairness has to be judged in each case in relation to a particular item of information. If different items of information are obtained separately, each must be fairly obtained.

Sometimes information is obtained from a source other than the data subject. In appropriate cases, regard is given to the fairness to the data subject as well as the source of the information.

3.3.2 First Principle second part

"Personal data shall be processed fairly and lawfully"

The Registrar's Guidelines state that fairness of processing is judged by reference to the purpose of the processing, the nature of the processing and its consequences for the individual affected by it. So, for example, it is unfair for a data user to process personal data with the result that unsolicited marketing material is sent to an individual who has informed the data user that he or she does not wish to receive such material.

Data users, particularly those setting up Web pages, which ask for persons to input personal data, therefore need to consider the use of "tick-boxes" to enable persons to select whether to allow their personal information to be passed or sold on to others (for example, for marketing purposes). Data users need to register under the DPA in the first place.

The question of how to interpret "fair processing" was raised in *Equifax Europe Limited v Data Protection Registrar* Data Protection Tribunal DA/90 25/49/7.

The case revolved partly around the issue of the definition of "processing". The definition contained in Section 1(7) of the DPA refers, in relation to personal data, to processing being "by reference to the data subject". In this case, data was dealt with by reference to house addresses, not individual names, and *Equifax* argued that the data was not handled by reference to any data subject. This argument was rejected. Where the object of the exercise was to gain information about an individual, even if the method of searching for such information was not based on the individual's name, processing was by reference to the data subject.

| 3.3.3 | Second Principle | *"Personal data shall be held for one or more specified and lawful purposes"* |

This principle would be contravened, for example, by an organisation whose registration merely extended to using personal data for internal personnel management purposes and who then used that data for marketing purposes.

The current application form for DPA registration asks for details of to whom the data held is to be disclosed. Considering the global nature of the Internet it may be that organisations going on-line need to amend their registration in order to account for the potential disclosure overseas; a simple box-ticking exercise.

| 3.3.4 | Third Principle | *"Personal data held for any purpose or purposes shall not be used or disclosed in any manner incompatible with that purpose or those purposes"* |

Part 2 of Schedule 1 to the DPA provides that personal data shall not be treated as used or disclosed in contravention of this Principle unless:

(a) the person to whom the disclosure is made is described in the disclosures section of the part of the data user's register entry which relates to those data, or

(b) the disclosure is made in circumstances covered by one of the "non-disclosure exemptions".

Page 61 of the current the DPA guidelines provides more information about this.

The word "incompatible" suggests that something more is required than mere compliance with these conditions. Indeed, the Registrar's Guidelines state that disclosures may be made or data used for a purposes not described in a data user's register entry without there necessarily being any incompatibility. Whether a use or disclosure is compatible with the registered purposes is a question of fact for the Registrar to consider on a case by case basis.

Note, however, the provisions of Section 5(2)of the DPA, which provide that it is an offence, *inter alia*, to hold or use personal data for any purpose other than the purposes described in the entry or to disclose personal data to anyone not described in the entry.

There has been an interesting recent development regarding the interpretation of the word "use". In February 1996 the House of Lords, in the case of Regina v Brown, affirmed an earlier Court of Appeal's decision and ruled that the retrieval and viewing on screen of personal data, without doing any further act with the information retrieved, is not "use" for the purposes of Section 5(2)(b) of the DPA. In other words the mere act of accessing the information stored in the computer and displaying it (which is regarded as processing of data) is not construed as "use", as long as nothing else is done with the information.

3.3.5   Fourth Principle

"Personal data held for any purpose or purposes shall be adequate, relevant and not excessive in relation to that purpose or those purposes"

Generally, the question of irrelevant information is fairly simple to determine. However, there may be problems in judging whether data is adequate or excessive in relation to the purposes for which it is held. There may also be a grey area between inadequate information and inaccurate information (see Fifth Principle). The Registrar has stated that the Fourth Principle should be read as a whole and that therefore the data should be judged on all three categories; in other words that the terms "adequate", "relevant" and "excessive" should be considered together so that data users identify the minimum amount of information they require to hold for their purposes.

3.3.6   Fifth Principle

"Personal data shall be accurate and, where necessary, kept up to date"

The question of whether or not personal data is accurate is determined in the same way as under DPA s.22. DPA s.22(4) provides that data is inaccurate if incorrect or misleading as to any matter of fact.

Under DPA s.22(2), data which accurately records information received or obtained by the data user from the data subject or a third party is to be treated as accurate for the purposes of the section (and therefore also this Principle) if the following have been complied with:

(a) the data indicates that the information was received or obtained as aforesaid or the information has not been extracted from the data except in a form which includes an indication to that effect, and

(b) the data subject has notified the data user that the information is regarded as incorrect or misleading, an indication to that effect has been included in the data or the information has not been extracted from the data except in a form which includes an indication to that effect.

DPA s.22(3) is not specifically mentioned in Part 2 of Schedule 1 dealing with this Principle, but there is an implication, reflected in the Registrar's Guidelines, that the obligation on data users under this Principle to ensure accuracy is not an absolute one. Instead, a defence may be the fact that a person has taken such care, as in all the circumstances was reasonably required, to ensure the accuracy of the data at the time.

This shows the dynamic nature of the requirement of the DPA and highlights the practical steps that must be taken by data users on the Internet. Updating and maintaining accuracy of information is crucial on the Internet and not merely restricted to data protection issues. Before allowing data to be put onto the Internet, a data user should consider issues such as:

- the significance of accuracy and whether the degree of accuracy has or is likely to cause damage or distress to the data subject

- the source from which the information was obtained and whether it was reasonable for the data user to rely on the information received from that source

- whether the data user tried to check the accuracy of the information with another source and whether it would have been reasonable to ask the data subject if the information was accurate

- the procedures for data entry and for ensuring that the system itself did not introduce inaccuracies into the data

- the procedures followed by the data user when the inaccuracy came to light including whether the data was corrected, whether the correction was passed on and what the data user has done about any consequences of an inaccuracy.

These issues may be considered by the Registrar.

This Principle requires that personal data should be kept up to date "where necessary". The Registrar's Guidelines state that the purpose for which the data is held or used is relevant in deciding whether updating is necessary. The Registrar may wish to take into account whether:

- any record is kept of the date on which the information is recorded or last updated

- those involved with the data are aware that it may not be current

- updating procedures are in place

- outdated information is likely to cause damage or distress to the data subject.

3.3.7   Sixth Principle

"Personal data held for any purpose or purposes shall not be kept for longer than is necessary for that purpose or those purposes"

The Registrar's Guidelines state that, in order to comply with this Principle, data users need to review personal data regularly and delete data which is no longer required for their purposes. It is suggested that a systematic review policy is instituted by data users.

Page 65 of the Registrar's Guidelines states that:

"If personal data has been recorded because of a relationship between the data user and the data subject, the need to keep the information may no longer exist when the relationship ceases to exist...".

However, it may be necessary to keep some information about the data subject; for example, to be able to defend legal claims which may be made in the future. In this case, when the relevant statutory time limit for making a claim has expired, the data should be deleted.

If all information is kept on-line then its electronic form facilitates manipulation and distortion by those inclined to do so. Security measures such as encryption (see 9.9.3) therefore play an important role in Internet usage.

Section 7.5 includes more detail on the weight of evidence and value of electronic documents.

Page 65 of the Registrar's Guidelines states that:

"The DPA provides that where personal data is held for historical, statistical or research purposes and is not used in such a way that damage or distress is or is likely to be caused to any data subject, the data may be kept indefinitely".

When considering procedures for deletion, the rights of individuals or the subject of personal data to compensation for damage and distress arising from loss or destruction of the data should be considered.

3.3.8   Seventh Principle

"An individual shall be entitled -

(a) at reasonable intervals and without undue delay or expense -

(i) to be informed by any data user whether he holds personal data of which that individual is a subject; and

(ii) to access any such data held by a data user; and

(b) where appropriate, to have such data corrected or erased"

Paragraph (a) of this Principle should not be construed as conferring any right inconsistent with Section 21 of the DPA.

Section 21 entitles individuals to be informed by any data user whether the data held by that data user includes personal data of which the individual is the data subject and to be supplied by the data user with a copy of the information constituting any such personal data.

However, the data user need only supply such information in response to a written request and on payment of a fee. A request to be informed whether any personal data is held should be treated also as a request to be supplied with the information making up the personal data.

Increasing use of the Internet could lead to more people and organisations requiring registration under the Act. This could, in turn, increase the likelihood of unlawful disclosure of data or for data to become incorrect. It is likely, therefore, that public awareness of data held on themselves will increase, as will requests to see that data.

This moves the discussion away from strictly legal considerations, to issues of human rights and civil liberties; the delicate balance between the freedom of the individual and control by the state.

Where a data user has separate entries in the Register in respect of data held for different purposes, a separate request must be made and a separate fee paid in respect of the data to which each entry relates.

A data user need not comply with the requests under Section 21 unless the person making the request can be identified sufficiently to look up the information sought.

The Registrar can enforce a subject's right to access under the Seventh Principle only when the Registrar is satisfied that the data user has contravened Section 21 of the DPA by failing to supply information to which the data subject is entitled which has been duly requested in accordance with that Section.

Page 66 of the Registrar's Guidelines states that:

"Correction or erasure of personal data is only "appropriate" when necessary to ensure compliance with the other Data Protection Principles".

Thus, the most likely Principles to be contravened are the Second Principle (data to be held only for one or more specified and lawful purposes), the Sixth Principle (data not to be kept longer than is necessary for the purpose or purposes for which it is held) and the First Principle (personal data to be fairly and lawfully obtained and processed). Correction and erasure is not appropriate merely on the ground that the data user has withheld subject access. The right to have data corrected or erased does not entitle a data subject to have personal data deleted merely because the data subject would prefer that the data user should not hold that information.

3.3.9   Eighth Principle

"Appropriate security measures shall be taken against unauthorised access to, or alteration, disclosure or destruction of, personal data and against accidental loss or destruction of personal data"

This Principle applies to personal data in respect of which personal services are provided by computer bureaux as well as to personal data held by data users. This principle seems to be the most relevant in implying security obligations upon those operating computer networks containing personal information.

The following are relevant:

- the nature of the personal data and the harm which would result from such access, alteration, disclosure, loss or destruction

- the place where the personal data is stored

- the security measures actually programmed into the relevant equipment (passwords for example)

- the measures taken for ensuring reliability of staff having access to the data. (Therefore guidelines for access are a relevant consideration: see 9.2.3).

The Registrar's Guidelines state that the prime responsibility for putting into practice a security policy rests with the computer user. The policy should seek to achieve that personal data can only be accessed, disclosed or destroyed by authorised people; that those people act within the scope of their authority; and that, if the data is accidentally lost or destroyed, it can be recovered so as to prevent any damage or distress being caused to the data subject. Questions which may need to be considered include:

- are disk files deleted before re-use or is new data merely written over the old?

- are printouts disposed of securely?

- is responsibility for an organisation's security policy clearly placed on a particular person or business function?

- are sufficient resources and facilities made available to enable that responsibility to be fulfilled?

- are breaches of security properly investigated and remedied?

- are responsibilities for security clearly defined between a computer bureau and its customers?

The mere fact that there has been a breach of security will not cause the Registrar to take formal action but the Registrar will want to be satisfied that the computer user has done everything which could reasonably be expected in order to avoid the breach.

Will employers who, for example, freely allow their employees unsupervised access to the Internet, be taken as having set up such appropriate security measures? Arguably no, because personal data may be disclosed, accessed, altered, etc. extremely easily in the context of the information society.

3.4   Exemptions

3.4.1   National security

Under Section 27 of the DPA, personal data is exempt from the provisions of the Act if an exemption is required for the purposes of safeguarding national security. Whether the exemption is or was required shall be determined by a Minister of the Crown whose certificate to that effect is conclusive evidence of that fact.

Even where this exemption does not apply, personal data may be exempt from non-disclosure provisions in cases where disclosure of the data is required to safeguard national security. Again, a certificate signed by a Minister of the Crown to this effect is conclusive evidence of this.

It is likely that a considerable amount of government-held information may be afforded such an exemption from disclosure.

3.4.2   Crime and taxation

Under Section 28 of the DPA, personal data held for:

> *"(a) the prevention or detection of crime;*
>
> *(b) the apprehension or prosecution of offenders; or*
>
> *(c) the assessment or collection of any tax or duty*
>
> *are exempt from the subject's access provisions in any case in which the application of those provisions to the data would be likely to prejudice any of the matters mentioned in this sub-section."*

Personal data held for the purpose of discharging statutory functions and consisting of information obtained for such a purpose from a person who had it in his or her possession for any of the purposes mentioned above, is exempt from the subject access provisions to the same extent as personal data held for any of those purposes.

3.5   Obligation to register

A person may not hold personal data unless an entry in respect of that person as a data user, or as a data user who also carries on a computer bureau, is contained in the register [see DPA s.5(1)].

In the case of a computer bureau, a person may not provide services in respect of personal data unless an entry in respect of that person as a person carrying on such a bureau, or as a data user who also carries on such a bureau, is contained in the register [see DPA s.5(4)].

Section 7(2) of the DPA states that an application for registration may be refused where the Registrar:

> *"(a) considers that the particulars proposed for registration will not give sufficient information as to the matters to which they relate, or*
>
> *(b) is satisfied that the applicant is likely to contravene any of the Data Protection Principles, or*

> *(c) considers that the information available to him is insufficient to have confidence that the applicant is unlikely to contravene any of those Principles."*

This does not stop the Registrar accepting particulars expressed in general terms in cases where that is appropriate. For example, the office of the Data Protection Registrar has been happy to accept highly tailored applications with regard to some multimedia products.

Where the Registrar refuses an application under Section 7(2), reasons must be given and applicants informed of their rights of appeal conferred by the DPA.

3.6 Offences and liabilities under the DPA

3.6.1 Processing personal data

The DPA has teeth. It specifies several offences in relation to the unlawful processing of personal data. These consist of:

(a) unregistered holding of personal data [DPA s.5(1)]

(b) non-compliance with the registered particulars [DPA s.5(2)], including:

- holding personal data of any description other than that specified in the entry

- holding any such data or using any such data for the purpose other than the purpose described in the entry

- obtaining such data, or information to be contained as such data, to be held from any source which is not described in the entry

- disclosing any such data to any person not described in the entry

- directly or indirectly transferring such data to any country or territory outside the UK other than one named or described in the entry.

3.6.2 Offences by computer bureaux

These consist of:

- knowingly or recklessly providing services in respect of personal data unless an entry in respect of the person carrying on such services as a person carrying on a computer bureau or as a data user also carrying on a computer bureau is registered

- knowingly or recklessly disclosing personal data in respect of which services are provided by a person carrying on a computer bureau without prior authority of the person to whom the services are provided.

3.6.3 Liability of officers

Where an offence under the DPA has been committed by a body corporate and is proved to be committed with a consent or connivance of director, manager, secretary or similar officer, that person as well as the body corporate shall be guilty of the offence. This is referred to at Section 20 of the Act.

3.6.4 Liability of servants and agents

Under Section 5(3) of the DPA, a servant or agent of a person to whom the obligation in relation to registered particulars [referred to at Section 5(2)] applies, shall be subject to the same restriction on the use, disclosure, or transfer of the data as the person registered. Contravention of the obligation of computer bureaux not to disclose personal data without the prior authority of the person for whom the services are provided also applies to servants and agents of the person carrying on the computer bureau.

3.6.5 Sanctions

The offences above are punishable as follows:

(a) on conviction or indictment, by a fine; or

(b) on summary conviction by a fine not exceeding statutory maximum (at present, £5,000).

3.7 Registrar's jurisdiction and powers of enforcement

3.7.1 Enforcement notices [DPA s.10]

The Registrar, if satisfied that a registered person has contravened or is contravening any of the Data Protection Principles, may serve an enforcement notice requiring that person, within a time specified in that notice, to take specified steps for complying with the Principle(s) in question.

Enforcement notices should contain a statement of the main Principles which the Registrar considers have been or are being contravened, with the reasons for reaching that conclusion and details of the right of appeal.

Any person failing to comply with an enforcement notice is guilty of an offence. It is however a defence to prove that one has exercised all due diligence to comply with the notice in question.

3.7.2 De-registration notices [DPA s.11]

The Registrar, if satisfied that a registered person has contravened or is contravening any of the Data Protection Principles, may serve a de-registration notice stating the intention, at the expiration of a period specified in the notice, to remove from the register all or any other particulars constituting the entry or any of the entries contained in the register in respect of that person. At the end of that period the Registrar may remove the particulars in question from the register. There is a right of appeal.

3.7.3 Transfer prohibition notices [DPA s.12]

If it appears to the Registrar that a person, registered as a data user and is a data user who also carries on the business of a computer bureau or is someone treated as so registered, proposes to transfer personal data held by that person to a place outside the UK, then the Registrar may serve that person with a notice prohibiting the transfer of the data either absolutely or until that person has taken such steps as are specified in the notice for protecting the interest of the data subjects in question. The question is how to protect those interests? Particularly if data is to be transferred to countries who may have lower standards of protection than the UK.

Such notices may be served when:

- the place to which the data is to be transferred is not a state bound by the Council of Europe Convention and the Registrar is satisfied that the transfer is likely to contravene or lead to a contravention of any of the Data Protection Principles

- the data is to be transferred to a state bound by the Council of Europe Convention and the Registrar is satisfied that either:

  (i) the person in question intends to give instructions for the further transfer of the data to a place which is not in such a state and that the further transfer is likely to contravene or lead to a contravention of any of the Data Protection Principles; or

  (ii) in the case of data to which an order under Section 23 of the DPA applies, the transfer is likely to contravene or lead to a contravention of any of the Data Protection Principles as they have effect in relation to such data. (Note that Section 23 applies to any data in relation to the racial origin of data subjects, their political opinions or religious or other beliefs, their physical or mental health or their sexual life, or their criminal convictions).

Contravention of a transfer prohibition notice is an offence unless the person charged proves that all due diligence has been exercised to avoid contravention.

3.7.4 Power of entry and inspection

Under Schedule 4, the Registrar is given the power to obtain and execute a warrant for entry to and inspection of premises where there are reasonable grounds for believing that an offence under the DPA has been or is being committed and to which access has been unreasonably denied after seven days written notice.

3.7.5 Limitation of Registrar's powers

It should be noted that the Registrar has no direct power to serve notices on unregistered persons; however, as mentioned above, there are a number of criminal offences under the DPA which are relevant to unregistered persons; most noticeably that of holding data without being registered.

The Registrar has no power to require information either from registered or unregistered persons.

3.7.6 Territorial extent of the DPA

As a general rule, the DPA does not apply to a data user in respect of data held, or to a person carrying on a computer bureau in respect of services provided outside the UK. Data is treated as held where the data user exercises control over the data.

Where persons who are not resident in the UK exercise this control through a servant or agent in the UK, the DPA shall apply to the servant or agent as though acting on that person's behalf. Such persons may be registered by reference to the position or office which they hold, and the entry shall apply to the occupant from time to time of the office or position.

The DPA does not apply to data processed wholly outside the UK unless intended for use, or used, in the UK.

3.8 Data Protection Directive implications for the Internet

The EU Directive on Data Protection was adopted on 24 October 1995, and is to be implemented by Member States by 24 October 1998. The Directive is concerned with the protection of individuals with regard to the processing of personal data and on the free movement of such data. The object is to harmonise data protection law in Member States so as to facilitate the free movement of personal data across frontiers, whilst protecting the essential rights of data subjects. The following guidance does not deal in detail with the provisions of the Directive; it merely provides an outline of the main points of note.

Processing definition

The Directive has extended the definition of "processing" from the DPA's present one of "amending, augmenting, deleting or re-arranging the data or extracting the information constituting the data" to include "any operation or set of operations performed upon, or any use, etc, of personal data whether or not by automatic means". This could lead to users of computers, through which Internet messages are passed, being held to process personal data. Arguably, they technically control the content and use of the data, and risk being considered data users. This must be viewed as unlikely because it is fairly tenuous legally and because of the practical impact of every person on the Internet being a data user.

Personal data definition   The definition of "personal data" in the DPA is extended to any information relating to an identifiable natural person (such a person need not be identifiable by the data user).

Transfers to third countries   Transfers to a third country of personal data which is undergoing processing or is intended for processing after transfer to the third country is only to be permitted if the third country ensures an adequate level of data protection. This could conflict with the intentions of organisations who wish to use the Internet in order to transfer such data.

Transfer of personal data to a third country which does not ensure an adequate level of protection may take place on condition that:

(a) the data subject has given unambiguous consent to the proposed transfer (potentially a massive task if there are a number of data subjects about whom data is to be transferred), or

(b) the transfer is necessary for the performance of a contract between the data subject and the controller (any person who alone or with others determines the purposes and means of processing personal data) or the implementation of precontractual measures taken in response to the data subject's request, or

(c) the transfer is necessary for the conclusion or performance of a contract made in the interest of the data subject between the controller and a third party, or

(d) the transfer is necessary on the grounds of public interest, or for the establishment, exercise or defence of legal claims, or

(e) the transfer is necessary in order to protect the vital interests of the data subject, or

(f) the transfer is made from a register which according to laws or regulations is intended to provide information to the public and is open to consultation either by the public in general or by any person who can demonstrate legitimate interest, to the extent that the conditions laid down in law for consultation are fulfilled in the particular case.

A Member State may authorise a transfer or a set of transfers of personal data to a third country which does not ensure an adequate level of data protection where the controller gives sufficient guarantees with respect to the protection of the privacy and fundamental rights and freedoms of individuals.

While the Directive extends the scope of protection afforded to data subjects, it should be noted that it is dependent on national implementation within the general framework set out by the Directive.

3.9 Implications of the DPA for the Internet

3.9.1 DPA definitions

As stated previously, at 3.2, whilst personal data itself must be recorded in a processable form, other information which enables one to identify the person need not be.

This is particularly pertinent for the Internet where a certain body of data may be held on one computer and other data on another. Individually, the data in both sites may not refer to living individuals, but together, they may identify an individual. Therefore the controllers of both sites may need to contemplate their relationship with the DPA.

The definition of a data user could potentially include the person running a bulletin board service.

The DPA applies if a person is a data user or computer bureau. Given the definition of "processing", potentially anyone using the Internet is a data user. A person is not a data user, however, in relation to data which passes through that person's computer en route to another, since the data is not being processed. This is also the case if a person is merely the passive recipient of data and does not process it. Therefore if an e-mail, potentially containing personal information, passes through a conduit terminal, the owner of the terminal is not breaching the provisions of the DPA.

Providers of electronic products and services are all likely to be data users; for example:

- on-line service providers dealing in personal data

- certain Web page holders, if "visitors" to the page are encouraged to leave details which may be circulated, or accessed by later "visitors"

- direct marketers

- credit reference agencies

- employment agencies

- travel agents

- financial services providers.

The above list is not intended to be exhaustive.

3.9.2   Offences under the DPA

As mentioned at 3.6.1, transferring data out of the UK is very likely in the Internet sense and therefore data users may need to amend their registrations.

The potential for personal liability of officers of a body corporate points to a requirement for clear internal guidelines for using the Internet.

3.9.3   Registrar's jurisdiction

As stated previously at 3.7.3, the Registrar has the power to issue a Transfer Prohibition Notice under Section 12 of the DPA.

The Internet, being global in nature, and particularly in view of the uncertain routes taken by communications, may make the issue of such notices become more common. Organisations would be wise to encrypt personal data for transfer over the Internet in order to protect the interests of data subjects.

The question of the territorial extent of the DPA (see 3.7.6) is a particularly important one in the context of the Internet where data may be transferred with great ease and there are potential problems with tracing the source of the information.

4 Public Records Act 1958

4.1 Introduction

The Public Records Act 1958 (PRA) was enacted in order to make provisions with respect to public records; that is provisions specifying what records are deemed to be public, what being "public" entails for those records and provisions specifying the period for which such records must be held.

The issue of what constitutes a "document" arises in a number of areas in relation to any discussions of the legal implications of doing business on the Internet and in the information society at large.

Digitisation allows text, images and sounds to be compressed, stored and transmitted over great distances in a relatively short time.

Other chapters discuss the evidential weight given to electronic images in legal proceedings (see 7.5.3) and the requirement for written documents under the law of contract (see 8.2).

The particular issue which prompts discussion of the PRA is the requirement to retain certain public records. To what extent does storage in electronic form cause problems in terms of authenticating the origin of public records? To what extent do government organisations need to archive electronic documents?

4.2 Synopsis of the PRA

4.2.1 The meaning of "records"

PRA s.10(1) states that:

"... "public records" has the meaning assigned to it by the First Schedule to this Act and "records" includes not only written records but records conveying information by any other means whatsoever".

It is clear, therefore, that records need not be written. What is required is that information is conveyed by some means.

Information converted into permanent form (for example, a printout) is considered as a record as is that which is saved to electronic storage (for example to a disk).

The question arises as to whether information on screen is sufficiently tangible to be a record and thus capable of being within the provisions of the PRA.

For information to appear on screen it must at some stage have been processed. However, when accessed, the information is copied onto the computer, even if only momentarily. In copyright, it is accepted that the act of loading or running software constitutes copying for the purposes of copyright law (see Chapter 1.10). Copyright subsists in the expression of an idea, whether that expression be in writing or otherwise.

By extension, an electronic work is still a work for the purposes of copyright law and so it is arguable that an electronic "public record" is still a "public record" if on screen. A screen display is "conveying information" and is a record (if only temporarily) of that information. In addition, for a document to be transmitted, it must have been stored first. Therefore, it seems very likely that the PRA applies fully to information which is transmitted over networks and must be considered when dealing with electronic information.

4.2.2   The meaning of "public"

"Public" records are defined in Schedule 1 paragraph 2 of the PRA which states:

"Generally, administrative and departmental records belonging to Her Majesty, Her Majesty's Government and any office, commission, other body or establishment under Her Majesty's Government are treated as public records".

Exactly which offices, commissions and other bodies and establishments apply is set out in a Table in Schedule 1 of the Act.

Other bodies referred to include The Meteorological Office, National Health Service Authorities, the National Insurance Advisory Committee and the Air Transport Advisory Council.

Other establishments referred to include the British Museum, Tate Gallery and the Royal Greenwich Observatory.

In addition, records of the Supreme Court and other courts and tribunals are treated as public records.

It is not clear whether the Schedule is updated from time to time to cover new agencies. At the same time there is provision for further categories of records to be added at the discretion of Her Majesty.

4.2.3   Offices established
        under the PRA

Section 1 establishes the general responsibility of the Lord Chancellor for Public Records. Section 1(2) sets up an advisory council on public records to advise the Lord Chancellor on matters concerning public records in general and in particular on those aspects of the work of the Public Records Office (PRO) which affects members of the public who make use of the facilities provided by the PRO. (The PRO is defined by Section 2 of the PRA as the repository for all records regulated under the PRA).

Section 3 establishes a duty of every person responsible for public records of any description which are not in the PRO or a place of deposit appointed by the Lord Chancellor under the PRA, to make arrangements for the selection of those records, which should be permanently preserved and for their safekeeping in the PRO. The Act imposes an obligation upon various bodies to put public records (as defined) into store at the PRO and then allow access to the public after a thirty year period.

4.2.4   Responsibilities

PRA Section 3 explains that every person responsible for public records in any government office has a duty to select and keep safe those records which should be preserved.

This means that end users/creators of new information have a positive obligation to assess and to send the records to the PRO. The Keeper of the Public Records is ultimately responsible for the collation and transfer of saved records to the PRO not later than 30 years after the records creation. (The Keeper of the Public Records is an officer appointed by the Lord Chancellor to take charge of the PRO and the records therein.)

4.2.5   Destruction

PRA Section 3(6) explains that, subject to the approval of the Lord Chancellor, records not selected for permanent preservation are destroyed, as are selected records which are duplicates of those already preserved.

4.2.6   Access

Records selected for the PRO are not revealed to the public until 30 years have elapsed, beginning with the first January following the records' creation. Obviously certain enactments can prevent public access even after the 30 year period. Similarly, access can be denied should inspection constitute a breach of good faith on the part of Government or on the part of the persons who obtained the information. This would occur if there was disclosure of information for which an undertaking was given, at the time of selection, that a longer period of storage would apply.

4.3   Implications for the Internet

The PRA is legislation passed a long time before current technology was perceived.

From examination of the provisions of the PRA, it seems clear that electronic documents made available over the Internet count as "records" within the PRA, and therefore need to be stored. Organisations must appreciate the ease with which electronic records can be tampered with or altered. The potential exists, more so than in a paper-based society.

In terms of what exactly should be stored, if a screen display constitutes a record for the purposes of the PRA, it needs to be kept.

However, a transient display may be too ephemeral to require storage, particularly if the information is changed a number of times before it reaches its final form. As the CIMTECH Document Management Yearbook 1995 observed:

"Uncertainty regarding the legal status of electronic records generally, and in particular digital images of existing paper documents, has long been perceived as a barrier to the uptake of digital document management. If you must keep the paper to be safe, so the argument goes, the paper is by default still the master and thus the whole concept of an integrated digital document management system is undermined."

Do electronic records need to be copied onto paper for storage purposes? To do so would defeat the object of having electronic records at all. Arguably, electronic records may be archived electronically and therefore it seems likely that the PRA does not require a paper copy (see 7.2).

5 Pornography

5.1 Introduction

On a practical level, those providing facilities, and those using and receiving on-line services, need to consider their potential liabilities for the transmission and possession of pornography under a number of UK statutes.

The statutes with direct impact in this area are the:

- Obscene Publications Act 1959, as amended by Schedule 9 of the Criminal Justice and Public Order Act 1994 (the 1994 Act)

- Protection of Children Act 1978 (as amended by the 1994 Act)

- Criminal Justice Act 1988

- Telecommunications Act 1984

- Indecent Displays (Control) Act 1981

- Computer Misuse Act 1990.

This chapter looks at the relevant sections of these acts as they could affect users of the Internet

There has been a great deal of concern expressed in the press, on the Internet, and by the public at large about the spread of computer pornography.

It is of particular concern that computer pornography is readily available to children and young teenagers. Whereas a newsagent or parent can supervise access to "top shelf" magazines or adult films, many parents are unaware of the hardcore nature of information available over the Net. Unlike soft pornography available in most newsagents, the Net carries hardcore pornography involving extreme violence, bestiality and paedophilia.

Concerns about such pornography are likely to result in increased police activity in connection with bulletin boards which advertise themselves as providers of this type of material.

In the US, a Carnegie Mellon study found 68 commercial "adult" computer bulletin boards located in 32 states with a repertory of "450,620 pornographic images, animations, and text files which had been downloaded by consumers 6,432,297 times".[18]

Merely possessing certain types of pornographic information can constitute an offence. In what is believed to be the first successful prosecution for possession of child pornography, a management consultant has received a heavy fine, been made to forfeit his computer equipment and, as a result, has been sacked from his job.[19]

5.2 Obscene Publications Act 1959

Section 1(3) of this Act makes it an offence to publish or distribute obscene material . However there is a defence in Section 2(5) of the Act which states that a person will not be convicted of an offence if previously he or she has not examined the material and has no reasonable cause to suspect that it is obscene. This may provide a defence for a network operator or Internet service provider but if they become aware that there is a likelihood of their facilities being used to distribute obscene material, and choose to do nothing, then they risk prosecution.

Section 1(1) of the Act states that the test of obscenity is whether the article, if taken as a whole, would tend to deprave and corrupt persons who are likely to read, see or hear it.

The article must be published. Publishing is defined widely in Section 1(3) of the Act by three distinct groups:

- "sells, lets on hire, gives or lends" where publication is to an individual

- "distributes, circulates", where publication is on a wider scale involving more than one person

- where a mere offer for sale or letting on hire constitutes publication.

---

[18] .net Magazine Issue 7 June 1995, p.10

[19] "Net surfer convicted for cache of child porn disks" – by Martin Lynch, Computing Online, 2 November 1995

Therefore, almost any information society activity involving obscene material can be counted as publication.

The penalties for anyone successfully prosecuted under the Act include fines and imprisonment (on conviction or indictment) for a term of up to three years [see s.2(1) of the Act].

The Act was amended by Schedule 9 of the 1994 Act so that publication is now defined to include the transmission of electronically stored data which, on resolution into human readable form, is obscene. Therefore the fact that an obscene image is digitised does not enable the publisher to escape liability. The word "transmits" is not well defined. It is not clear whether, for example, transmission can occur entirely within a single computer system; if not, there could be an argument that the entire Internet is a single "system". Similarly, it is not clear whether the person who "transmits" the data is the one who had the material before the transmission, or whether it is the one who causes the transmission to happen.

5.3 Protection of Children Act 1978

Section 1 of this Act (as amended by Section 84 of the 1994 Act) makes it an offence to take, make, permit to be taken, distribute, show, possess intending to distribute or show, or publish any indecent photograph or indecent pseudo-photograph of a child.

"Pseudo-photograph" was introduced by the 1994 Act and amends the definition of photograph to now include:

"data stored on computer disk or by other electronic means which is capable of conversion into a photograph."

The definition of "pseudo photograph" had been added to Section 7 of the Act by Section 84(3) of the 1994 Act and means

"an image whether made by computer graphics or otherwise, howsoever, which appears to be a photograph."

Further, if the impression conveyed by the pseudo-photograph is one which is difficult to classify as either an adult or a child, but the predominant impression is that the person shown is a child then it shall be treated as such. This is intended to capture computer-generated and manipulated images.

Both a person or company may be charged with an offence under this Act and the penalties are very similar to those under the Obscene Publications Act 1959.

5.4	**Criminal Justice Act 1988**	Section 160 of this Act regulates the area of child pornography. The Act makes provision for a summary offence of possession of an indecent photo of a child.

Section 84(4) of the 1994 Act has amended the Criminal Justice Act 1988 in a similar way to the amendments to the Protection of Children Act 1978 by adding the term "pseudo photograph". |
| 5.5 | **Telecommunications Act 1984** | Section 43(1)(a) of this Act provides that it is an offence to send any message by telephone originating in the UK, which is grossly offensive or of an indecent, obscene or menacing character. The implications to the Internet of this subject are discussed at 5.8. |
| 5.6 | **Indecent Displays (Control) Act 1981** | Section 1 of this Act makes a person guilty of an offence if he or she publicly displays indecent material. Both the person making the display, and any person causing or permitting the display, are liable for prosecution.

Section 1(2) of the Act states that for material to be displayed, it must be visible from any public place .

The Act specifies the format of any warning to be used, and once again bodies corporate may face liability as well as individuals. |
| 5.7 | **Computer Misuse Act 1990** | Section 1 of this Act creates an offence of knowingly causing a computer to perform any function with intent to secure unauthorised access to any program or data held in any computer. If an Internet service provider makes it a condition of service that nothing unlawful is done, then downloading illegal material would become unauthorised access, and this Act could apply. |

**5.8   Implications of the
        law for the Internet**

There are quite a range of statutes currently in force covering the areas of pornography on the Internet. Two of these are being used in a current case in the UK which was reported in .net Magazine during 1995 as follows:

"Read all about it, Net porn update

The first Internet pornography case continues with the defendants Stephen Arnold and Alban Fellows remanded on bail to appear at Birmingham Magistrates Court on 20 April. They are charged with 18 offences under the Protection of Children Act 1978 and the Obscene Publications Act 1959."

It is clear the laws on obscenity do apply to new technology and have been amended where necessary to explicitly say so.

As with defamatory material, the defence under the Obscene Publications Act 1959 for persons who do not examine material to discover pornography is likely only to be available to the service provider and is not likely to be available to the user.

Once again it seems that the best advice for a service provider may be not to attempt to exercise any control over the service by way of monitoring content. However, looking closely at the wording of the defence under the Obscene Publications Act 1959, "did not examine, *and had no reasonable cause to suspect"*....... (emphasis added) it is clear that mere inactivity is not a sufficient defence, in that if a service provider becomes aware that obscene material is on the service then the problem cannot be ignored. To ignore the problem would expose the service provider to the risk of prosecution.

The publishing groups referred to at 5.2 means that almost any activity by the information society involving obscene material can be counted as publication and can make the publisher potentially liable under the Obscene Publications Act 1959.

The provision of the Telecommunications Act 1984, referred to at 5.5, extends to data transmitted by a telephone line and therefore covers the use of the Internet. However, the ambit of the Act is intended to apply to the originator of the material rather than the person distributing it. Therefore, it is unlikely for a service provider to be caught by this provision in the Act but it applies to the originator of the material.

The definition of display under the Indecent Displays (Control) Act 1981 includes the Internet and, for example, terminals in public libraries, "cyber-cafes", etc. However, Section 1(3) of the Act makes it clear that payment of a fee to view the material has the effect of making that material not on public display. Hence, a Web site, accessed only via a subscription mechanism or an adult bulletin board with similar pay-access, is not covered.

A perennial problem of the Internet is caused by its global nature; the conflict when one has to consider which laws should apply to the content of information held on the Internet.

A recent case in the US (the *Thomas* case) concerned a bulletin board operator specialising in sexually explicit material. The bulletin board was sited in California (whose rules on the subject are relatively relaxed) but the accused was prosecuted under Memphis law (considerably more strict), Memphis being the place where the information was accessed.

Should people who post pornographic pictures from the UK to New York-based Usenet groups be held liable in New York given that they had no way of knowing where the images might be downloaded? Would they be liable if children downloaded the images? Would the New York network operators be liable for importing obscene material? The law on these questions of liability is still obscure.

5.9   US developments

Public concern has recently caused the US Senate to pass, by an overwhelming majority, a bill known as the Exon Bill (after its author) or the Communications Decency Act, which gives the Federal Communications Commission the power to regulate "indecency" on the Internet. However, the Bill is being challenged under administrative law as critics feel it contravenes First Amendment civil rights. Internet enthusiasts and system operators also argue that the Exon Bill is unconstitutional and unworkable.

On 5th March 1996 Senators Leahy and Goodlatte proposed the "Encrypted Communications Privacy Act", which has been described as a very important step towards securing privacy and confidentiality for users of all new media. This bill is very much "anti-Exon" and it will be interesting to see how these two pieces of legislation square up to one another.

Because of the First Amendment, the US has probably the world's most highly developed case law on what constitutes obscenity. However, US law interprets obscenity by "community standards" and as the *Thomas* case shows (see 5.8), such standards vary considerably between communities.

One development towards censorship was the decision in December 1995 by CompuServe, one of the large American commercial networks, through which users can exchange information, to cut off its customers' access to more than 200 newsgroups on the Internet because of their pornographic nature.

The laws of cyberspace are particularly frontier-like in this area. Case law may shake down a sense of order. Whatever the fate of this recent legislation and the various court cases, the concerns raised will not go away. Battles over on-line freedom of speech and freedom of expression will continue to rage.

6   Regulatory environment

6.1   Introduction

There are those who would claim that cyberspace is lawless. Because the Internet transcends national boundaries and legal jurisdictions, so the argument goes, terrestrial laws cannot apply to cyberspace.

This is completely incorrect. The Internet is governed by all the normal rules; it is just their application which may cause difficulties.

In addition to rules of law, there are a great many regulations, particularly in industry sectors, which need to be complied with.

Technological developments change the social framework which the law is designed to regulate. Interactivity, coupled with media influence may become a particularly potent force in this respect. It is likely to erode some of the traditional legal boundaries which currently exist between work and leisure, and between workplace and home. Similarly, increasing interconnection could erode the distinction between public and private communication. It remains to be seen how far and how fast these trends unfold, but for long term projects their impact may be substantial.

The choice of services offered on the Internet will determine the regulatory regime applicable. By way of example, a cable TV station may wish to offer broadcasts which, amongst the possible programming, offer an interactive gambling facility. The regulation of such new services is discussed at 6.5.

This chapter examines a variety of regulations, concentrating on the telecommunications, advertising, and procurement sectors.

6.2   Internet culture and the need for regulation

The legal picture in cyberspace is very confused, for a number of reasons. Firstly, the sheer number of laws, in itself, can create a great deal of confusion. Secondly, there can be a number of different laws for any one activity, perhaps with conflicting requirements. Thirdly, laws conflict between countries; in the US there is even conflict between federal and state laws.

Many everyday activities find a new lease of life on the Internet. For example, a physical shop window can become a virtual one in a virtual shopping mall. Normal rules of contract and advertising apply but the great benefit of the Internet is its global nature and hence national laws have to be seen in the global context.

Once again, we return to the dilemma raised in 5.8; should the law of the country where the information resides on a computer apply, or the law of the country where that information was written or put onto the Internet, or where the information is accessed? There is no universally acceptable answer to this question but the issues are examined further at 10.2.

Sometimes the problem in emerging markets is that *de facto* standards can appear quickly and be monopolised by their developers. These developers feel no need to support their standard by licensing or pooling it, and so other developers are excluded. For example, in the computer games hardware market, standards are substantially set by SEGA and Nintendo, and are then protected as proprietary. In computing, a *de facto* standard is the combination of running Microsoft's "Windows" using Intel processors which represents 80% of personal computers sold[20]. In such cases potential competitors may seek to review their options under applicable pro-competition provisions in relevant intellectual property statutes, working against the standard holders.

These options include, in appropriate circumstances, the right to decompile software (including operating system software) in order to create interoperable software, and the opportunity to seek compulsory licences. An aggrieved company can also invite the competent EU and/or national authorities to investigate whether a standard holder is in breach of relevant competition principles law.

The EU has significant powers in this respect and is increasingly willing to define markets very narrowly and

---

[20] "The Software Revolution" by Amy Cortese, John Verity, Kathy Rebello and Rob Hof - Business Week, December 4 1995, p.44

so make a finding of dominance. The mere threat of the substantial fines which the EU can impose for abuse of dominant position and for anti-competitive agreements is a very effective tool for new entrants, and must be taken very seriously by those with a perceived entrenched advantage in the market.

6.3   Existing applicable regulations

It would be trite and of limited use to attempt to identify all the regulations applicable to the Internet. Other Chapters pick out major legal areas having a particular bearing.

Nevertheless, an Internet user needs to have an awareness for the sorts of laws, rules and regulations that can apply. For example, a manufacturer or distributor of an on-line electronic product or service (for example Web server) needs to know what line to take in order to conform to the law. The following may, for example, impinge upon the manufacture and distribution of multimedia products:

- Broadcasting Acts 1990 and 1995

- Competition Act 1980

- Consumer Protection Act 1987

- Copyright, Designs and Patents Act 1988

- Data Protection Act 1984

- Fair Trading Act 1973

- Health and Safety at Work (etc) Act 1974

- Patents Act 1977

- Resale Prices Act 1976

- Restrictive Trade Practices Act 1976

- Telecommunications Act 1984

- Video Recordings Act 1984

By way of example of regulatory difficulties, one issue recently topical in the UK was the extent to which the existing certification applied to movies might be required for interactive video games. Such a question arose from the wording of the Video Recordings Act 1984, which applies to video films.

The games manufacturer SEGA broke new ground by submitting to the British Board of Film Classification (BBFC) separately from the rest of their product range, the video footage used in one of its games.

The Times of Friday 25 June 1993 reported:

"SEGA's decision to submit its latest offering; Night Trap, to the BBFC is powerful evidence of the growing sophistication of video game technology".

This example shows, albeit in a specific field and jurisdiction, the conflict between the new opportunities offered by new technology and the existing industry regulations. For many organisations, taking the lead in setting industry standards will be seen as a public relations coup.[21]

6.3.1   Sources of law

There are various regulations covering Financial Services which have been mandated from the Financial Services Act 1986 and similar Acts.

As well as general law, (encapsulated by individual statutes and principles of common law) there are laws which apply to specific sectors or industries. A further category consists of "law" made by professional organisations or trade associations; for example, the Law Society rules of professional conduct regulate the conduct of solicitors. Such "codes of practice" are, in general terms, statements of good practice and, unless deriving from statute, are unlikely to have legal effect although they have weight in their particular industry sector.

6.3.2   Example codes of conduct

Rule 5(b)(I) of the Uncertificated Securities' Regulations 1992[22], which dealt with the TAURUS system, required systems upon which Stocks and Shares information was transferred to implement appropriate safeguards to reduce the possibility of error or fraud and unauthorised access to or manipulation of data.

---

[21] Interestingly, in the SEGA example, the BBFC assumed the power to regulate and deemed the game unsuitable for those under the age of 15 years

[22] S.I. 1992, No. 225

The Civil Service Management Code states, at Section 4.2, that:

"Departments and agencies must remind staff ... that they are bound by the provisions of the criminal law, including the Official Secrets Acts, which protect certain categories of official information, and by their duty of confidentiality owed to the Crown ..."

Rule 8 of the British Computer Society's Code of Conduct requires that:

"members shall not disclose, or use for personal gain or to benefit a third party, confidential information acquired in the course of professional practice, except with prior written permission of the employer or client, or at the direction of a court of law."

A common theme therefore appears to be the imposition of a duty upon members of a trade or industry sector, particularly with regard to valuable information. Transfer of information over the Internet puts the integrity of such information at risk, yet at the same time it is precisely that information which must be transferred for the information society to function effectively.

6.4   Telecommunications regulatory environment

The UK telecommunications regime has evolved over several decades. Organisationally the process started in the late 1950s when post and telecommunications began to be separated within the Post Office.

In the 1980s the process of change moved more quickly. In 1981 British Telecom was set up as a separate organisation. The British Standards Institution (BSI) and British Approvals Board for Telecommunications (BABT) progressively took over the standards and approvals work from British Telecom (BT). Progress was further accelerated when the Office of Telecommunications (OFTEL) was set up by the Telecommunications Act (TA) 1984 to be the regulator for telecommunications. This Act represented a major conceptual change. Hitherto, liberalisation and competition had been regarded as the exception; under the terms of the Act they became the norm.

Within the regulatory framework there are a number of organisations in which users can play a part. Although BABT is primarily concerned with suppliers, users are represented on its governing body. Users are also involved with the BSI's telecommunications standards committees. However, in each case representation is indirect being through user associations. Relationships with OFTEL can be more direct; it receives evidence from associations, but users can submit views on topics directly, and can make complaints about service or malpractice. Under the auspices of the Director General of Telecommunications (DGT), the contract terms for Public Telecommunications Operator (PTO) supply have been reformed, and a coherent tariff formula introduced and applied.[23]

In addition to the telecommunications implications for Internet service providers, the companies who buy bandwidth from the telecommunications operators, the Internet also allows voice telephony. The question therefore is whether an Internet service provider requires a licence under the TA to provide such a service.

Under Section 4 of the TA, a telecommunications system is defined as a system through which speech, music, visual images and various other signals pass. This would seem to cover the Internet. The global nature of the Internet means that the exemptions from licensing under Section 6 of the TA do not apply and so an Internet service provider offering Internet telephony is required to obtain a licence under Section 7 of the TA.

As telecommunications, broadcasting and computing markets merge, many existing regulatory frameworks are becoming inadequate. No single regulatory body or set of regulations expressly governs the information society. Nevertheless, all players must consider their position under a variety of different statutes and rules.

The Council of the European Commission agreed recently to set out a timetable for the liberalisation of telephone services in the EU. The main objectives include

---

[23] "Telecoms Users Guide to Regulations", Chapter 1, Section 1.9, p.1

creating a liberal environment for the international provision of telecom services, developing a structure within which to operate interconnection agreements, giving independence to telecommunications organisations, standardising the definition of universal service principles and consolidating the current regulatory environments.

OFTEL believes the market will tend towards development of broadband switched mass-market services and are keen to avoid both over-regulation and regulation where it is not needed.

OFTEL issued a consultative document on the regulation of broadband switched mass-market services delivered by telecommunication systems entitled *"Beyond the telephone, the television and the PC"*. The consultation period ran until 30 November 1995 and comments were invited on a variety of issues including questions of market dominance, standardisation and arrangements with content providers.

The OFTEL consultative document goes further than the TA. It queries the extent to which the owners of distribution networks should be willing to accept any service provider onto their networks. On the cable side, the European Commission has adopted a Directive (Cable Directive 95/51/EC) which allows European cable operators to offer telephony and other telecommunications services thus giving them much the same freedom as their UK counterparts. The directive means that cable operators are allowed to offer telecom services, although at the same time it seeks to restrict telecoms operators from owning telecoms and cable television infrastructures in the same geographical area.

Such developments have been welcomed by the European Cable Communications Association. The Directorate General IV (DGIV) sees this EU directive as mirroring the development of the information superhighway in the US. As such the directorate is looking at pan-European network and service projects and the regulatory barriers to the creation of a pan-European network.

6.5 Regulation of new services

As the Internet develops to offer new services, and provide existing services in new ways, the choice of services available will determine the regulatory regime applicable. By way of example, a cable TV station may wish to offer broadcasts which offer interactive gambling facilities.

In order to play, viewers would have to make a wager on the telephone or through a pay-per-view chart. In the US such a set-up must comply with the Federal Communications Commission (FCC) rules against broadcasting games of "chance" and the service provider must offer programmes that the regulators consider are based on "skill". Similarly, in the UK the service provider would need to adhere to the rules under the Betting, Gaming and Lotteries Act 1963, Gaming Act 1968, Lotteries and Amusements Act 1976 and the Betting and Gaming Duties Act 1981.

Regulations under these Acts include the need for a Betting Office licence, a bookmaker's permit and the necessity of paying betting duty. In addition it is important to note that betting by those under 18 is a criminal offence. There are also strict codes concerning advertising and betting which effectively restrict it being advertised on certain media; especially TV and Radio. This example shows the myriad of regulations which may apply to a particular activity.

6.6 Advertising

The "British Code of Advertising Practice" exists to increase public confidence in advertising and was set up by the Advertising Standards Authority (ASA). The Web is essentially a new marketing tool affording more ways to communicate more information.

The Code sets out the rules, drawn up by the advertising industry itself, on what is acceptable in advertisements, sales promotions and direct marketing. Note also that the Office of the Data Protection Registrar issued guidance in 1995 to deal with direct marketing[24]. It also covers all print, post, direct mail, unsolicited mail (including

---

[24] "Data Protection Guidance for Direct Marketers" - an Office of the Data Protection Registrar brochure, October 1995

e-mail), viewdata and cinema advertising, although not radio or TV commercials, which are the responsibility of the Radio Authority (RA) and the Independent Television Commission (ITC), together with the Independent Television Association.

If an advertisement breaks ASA rules, the authority can ask the advertiser to withdraw or amend it. Other sanctions include; adverse publicity, the refusal of further advertising space, removal of trade incentives and, finally, legal proceedings via a referral from the ASA to the Office of Fair Trading (OFT). The OFT has the power to obtain an injunction against advertisers to prevent them from repeating the same or similar claims in future advertisements.

The law imposes certain requirements with respect to advertising and marketing information. The Misrepresentation Act 1967 prohibits the making of untrue statements which induce a person to make a purchase. The deliberate use of a false description when offering to supply particular goods is an offence under the Trade Descriptions Act 1968.

In 1995, the airline Virgin Atlantic was fined in the US for inadvertently advertising incorrect fares on their Internet Web site. Under English law, such a situation would have been referred to the ASA, as a breach of the "British Code of Advertising Practice". In extreme cases, a reference could be made to the Director General of Fair Trading, under the Fair Trading Act 1973.

6.7 **Procurement regulations**

Procurement regulations are UK Regulations which implement the EU procurement rules into UK law in the fields of supplies and services. The UK Regulations are as follows:

- Public Supply Contracts Regulations 1995, SI 1995 no. 201, which took effect on 21 February 1995

- Public Services Contracts Regulations 1993, SI 1993 no. 3228, which came into force on 13 January 1994

- Utilities Supply and Works Contracts Regulations 1992, SI 1992 no. 3279, which came into force on 13 January 1993.

These Regulations were intended to enact in their entirety, and without change, the EU Directives relating to supplies and services. Unfortunately, this does not mean that the Regulations entirely restate the Directives. Therefore, conflict can and does arise. In the event of conflict, the EU Directives prevail.

The UK Regulations are based upon fundamentals of EU law. The Directives, upon which the Regulations are based, are designed to supplement Article 30 of the Treaty of Rome which prohibits restrictions on the free movement of goods[25].

Particular Internet related issues arise in at least two situations:

- the publication of notices in the Official Journal of the European Communities

- giving information to suppliers.

Both of these situations are discussed in more detail in Sections 4.2 and 4.5 respectively of Chapter 4 of the *Guideline* that is a companion to this *Reference Book*.

6.8   Other regulations

At present there are no general statutory regulations restricting the type of medium to be used for communication and record-storage. However, specific regulations may impose requirements and have certain implications which need to be considered such as:

- certain information may need to be incorporated into any communication as required by the Companies Act 1989

- a regulatory authority may have statutory powers to investigate and audit an organisation's electronic systems (for example the Finance Act 1985)

- organisations may have regulatory obligations to communicate information to public authorities and, increasingly, such regulations expressly permit the submission of such information in electronic form; for example, the Statistics of Trade (Customs and Excise) Regulations 1992.

---

[25] The term "supplies" is now employed in general everyday use rather than "goods"

7 Evidential issues

7.1 Introduction

This chapter provides a summary of English law on evidence as it currently stands; it is not an attempt at a comprehensive description of the laws. This is because the rules of evidence are complex and differ much in the same way as between civil and criminal proceedings and between various courts[26]. This chapter does, however, touch on those aspects of the rules that have particular relevance to electronic records and seeks to draw some general conclusions.

A new Civil Evidence Act received Royal Assent on 8 November 1995. When it comes into force this new legislation should make the admission of electronic records in court substantially easier. For instance, it abolishes the prohibition on admitting hearsay evidence. However, the precise implications of the Act depend on rules to be issued under the Act, which are not expected to be drafted until some time in 1996. The likely effects of the Act are summarised in 7.3, but for the moment the existing law continues to apply.

7.2 Current law

Under the existing law an electronic record is admissible if:

- it constitutes "real" evidence; or

- it constitutes "hearsay" evidence, but is admissible under an exception to the general rule prohibiting hearsay evidence

and in some cases[27] it satisfies the special rules governing the admissibility of computer-produced documents described below.

Real evidence in this context is an electronic record whose content did not originate in a human mind.

---

[26] Magistrates' courts, for instance, are governed by the Evidence Act 1938. We do not describe that here.

[27] See *Regina v Shepherd* [1993] 1 A.E.R. 225. This House of Lords case decided that the provisions of s.69 Police and Criminal Evidence Act 1984 apply to both real and hearsay evidence.

An example of real evidence is a date or time stamp on a computer file, taken directly from the system clock. If the time stamp originates from a date keyed in by an operator, then the record is hearsay. In either case, if there is a dispute about the date or time of the message, the court needs to know firstly whether it can look at the electronic record and, if it can, whether the record is reliable evidence of the date or time.

In the case of real evidence there may have to be "foundation testimony" which describes how the date stamp is derived. This may have to extend to the reliability of the system clock (is it checked regularly, is it changed when the clocks go back, which system clock does the time stamp work from, local PC or server, and so on). In the case of a keyed-in date, the court has to give whatever weight it thinks appropriate in the absence of evidence from the person who keyed it in.

Section 69 of the Police and Criminal Evidence Act 1984 and Section 5 of the Civil Evidence Act 1968, provide mechanisms for the admission in evidence of computer-produced documents.

Section 69 of the Police and Criminal Evidence Act 1984 states that a statement in a document produced by a computer shall not be admissible as evidence of any fact stated therein unless it is shown that:

- there are no reasonable grounds for believing that the statement is inaccurate because of improper use of the computer

- that, at all material times, the computer was operating properly or, if not, that any respect in which it was not operating properly or was out of operation was not such as to affect the production of the document or the accuracy of its contents

- any relevant conditions specified in rules of the court are satisfied (so far no rules have been made).

Oral evidence of the operation of the system can be given by someone familiar with the operation of the computer in the sense of knowing what the computer is required to do and who can say that it is doing it properly.

In most cases this need not be a computer expert. In *Regina v Shepherd* evidence given by a store detective of the operation of the computer system that produced the till roll was held to be adequate. An alternative way of complying with Section 69 is for a person in a responsible position in relation to the computer to give a certificate under Schedule 3 of the Act.

Section 5 of the Civil Evidence Act 1968 is more complex. It provides that a statement contained in a document produced by a computer is admissible as evidence of any facts stated therein of which direct oral evidence would be admissible, if certain conditions are fulfilled. Briefly, the conditions are that:

- the document containing the statement was produced during a period when the computer was used regularly to store or process information for the purposes of any activities regularly carried on over that period

- over the period there was regularly supplied to the computer in the ordinary course of activities information of the kind contained in the statement

- through the material part of the period, the computer was operating properly

- the information contained in the statement reproduces or is derived from information supplied to the computer in the ordinary course of those activities.

There is a difference in principle between the criminal and civil provisions. Section 69 of the Police and Criminal Evidence Act 1984 has to be complied with whether the evidence in question is real or hearsay but, whilst the provisions of Section 5 of the Civil Evidence Act 1968 apply to hearsay evidence, it is not clear whether they apply to real evidence as well. However, the general reasoning of *Regina v Shepherd* seems to apply as much to civil cases as to criminal cases; that is, to civil evidence as much as criminal evidence. So it seems that Section 5 is also likely to apply to both real and hearsay evidence.

If a document picks up the time and date from the system clock, it is "information supplied to the computer" by virtue of Section 5. Indeed, Section 5 expressly states that human intervention is not necessary. However, it should not be forgotten that even if Section 5 does not have to be complied with in the case of real evidence, foundation testimony will be necessary. A good example of this is *Regina v Cochrane*, in which there was insufficient evidence of the workings of a building society cash machine and the mainframe computer to which it was linked to enable the court even to decide whether the till roll of the cash machine was real or hearsay evidence.

One additional aspect of civil evidence is that it may be possible to achieve the admission of a hearsay statement in a computer-produced document without complying with Section 5 of the Civil Evidence Act 1968. For instance it may be possible to admit hearsay contained in computer-produced documents under Section 2 of the Civil Evidence Act 1968 without complying with Section 5. Section 2 allows, in some circumstances, the admission of "first hand hearsay"; for example, a statement in a document written by someone with personal knowledge of the facts set out in the statement.

Several of the cases that have come before the courts have involved computer printouts containing evidence of dates and times. Something ostensibly that simple has caused major problems, even in the context of computer systems wholly under the control of one of the affected parties. How much greater are the problems when, as with public networks, the information recorded in electronic messages may be derived from other computers around the world. Take the typical header on an item of Internet e-mail. This contains information about computers through which the message has been routed. If any of that information is crucial, who is going to provide the necessary testimony to establish the reliability of that computer so as to support the admissibility of the electronic record?

7.3   Civil Evidence Act 1995

The Civil Evidence Act 1995 revolutionises the law of civil evidence. It abolishes the prohibition on admitting hearsay evidence and abolishes the special rules in Section 5 of the Civil Evidence Act 1968. However, this does not mean that electronic records can suddenly be taken at face value. The same sort of evidence that was previously necessary to found the admission of real evidence or to comply with the rules for computer-generated documents is still needed to establish the weight and reliability of the record; whether it is real evidence or hearsay.

There are still special rules covering the mode of admitting hearsay evidence. The judge may be able to look at the electronic record if it is hearsay but still needs to be persuaded to place reliance on it. The new Act states that, in estimating the weight (if any) to be given to hearsay evidence in civil proceedings, the court shall have regard to any circumstances from which any inference can reasonably be drawn as to the reliability or otherwise of the evidence.

The new Act makes specific provision for admission in evidence of documents (ie. anything in which information of any description is recorded) shown to form part of the records of a business or public authority. Such documents can be admitted without further proof. A document is taken to form part of the records of a business or public authority if an officer of the business or public authority to whom the records belong signs a certificate to that effect.

"Officer" includes any person occupying a responsible position in relation to the relevant activities of the business or public authority or in relation to its records. Public authority includes any public or statutory undertaking or any government organisation and any person holding office under Her Majesty. The court may, having regard to the circumstances of the case, direct that these provisions or any of them do not apply in relation to a particular document or record or description of documents or records.

The full implications of the new Act will become apparent only when the various rules of court are made under it.

7.4	**Improving evidential quality of electronic records and communications**	There are various actions that can be taken to assist in improving the evidential quality, and the authentication and validation, of electronic records and communications. However, it is a complex area and the actions suggested are not comprehensive; and advice should be sought as to their proper use and degree of effectiveness.
	Maintain good procedures	A reliable person should be able to come to court and give credible testimony from personal knowledge as to how the system works, the checks carried out on it and the steps taken to ensure that the information (such as system clocks) on it is accurate and up to date.
	Use hard copies	A hard copy printed by the system and certified, dated and signed by the person who prints it off should be persuasive evidence of the state of the electronic record at the time it was printed. This may only be practicable in limited circumstances.
	Archive to CD-ROM	A well-maintained optical disk document archive can potentially form a reliable collection of snapshots of the state of the records on the system. The maintenance of such archives is a subject in its own right on which much work has been done in developing codes of practice.
	Use hash totals	Software is available that calculates a hash total of information to be transmitted. If the document is tampered with, the hash total changes. This allows tampering to be detected. Conversely, the fact that a hash total remains the same should be good evidence that the content of the electronic record has not changed. The quality of hash totals varies with their length (for example 160-bit is better than 16-bit), and the quality of the hashing algorithm. Of course, the hashing process may itself require testimony to explain it to the judge.
	Use digital signatures	This encryption technique enables a message to be tagged with a unique identifier which can be recognised by the recipient. This should identify the sender, so long as the private key used to generate the digital signature has not been compromised and the public key used to check the digital signature can be trusted. For instance, if a confirmatory response is required from the counterparty signed with a digital signature, the counterparty may have difficulty convincing a judge that it was not sent (see also 9.9.4).

7.5 Evidential issues and the Internet

7.5.1 Why evidential issues matter

If there is a dispute involving communications over the Internet, the court may require answers to questions such as:

- who sent the message?
- when was it sent?
- from which country was it sent?
- what route did the message take?
- what were the contents of the message?
- in which country was it received?
- when was it received?
- when was it read?
- who read it?

The answers to these questions can crucially affect issues such as whether and when a contract was made over the Internet, which country's laws and jurisdiction cover the contract, who authored, published or distributed defamatory or infringing material and so on.

7.5.2 Admissibility of electronic documents

The court has two main sources of evidence from which to decide the issues. First there is oral evidence from the people concerned, but whose evidence may of course conflict. Second there are electronic records and hard copy printouts of the messages and any logs recording the sending and receipt of the messages.

Contemporaneous documents often provide the key to resolving conflicts of evidence between witnesses. But this only happens if the records are ruled as admissible by the judge.

7.5.3 Weight of evidence

Where electronic records are ruled admissible, there still remains the question of weight. How much reliance will the judge place on the content of the record or of the log? If the record displays a time of sending, will the judge accept that this is strong evidence that the message was sent at that time or place little or no reliance on it? If the judge disregards the time record, will he nevertheless accept that the record accurately records the content of the message? It is well known that Internet Protocol (IP) addresses can be "spoofed" so that the true identity of the sender is disguised, therefore will the judge be convinced merely by the electronic record of the apparent sender?

These are the kind of issues to which rules of evidence are directed. For many communications on the Internet, it does not matter a jot whether the message is good evidence because nothing of any significance turns on the message. But for any messages that have contractual implications, or deal with matters of importance or value, or may incur liability, doubts about evidential value can create a real obstacle to using informal electronic communications. For these reasons, trading electronically typically takes place on specially constructed EDI networks, using a combination of formal technical and contractual structures to address these issues.

Currently, the safe course has to be proving the message was sent and received by the correct persons, and proving the content of the message, unless there is in place some reliable means of identifying the other person in the transaction. Without this evidence it is unsafe to conduct significant transactions or business on the Internet. Mechanisms such as digital signatures, designed to overcome problems of identification, are discussed in 9.9.4.

| 7.5.4 | Which laws might apply |

If a person has the relevant electronic records of an Internet communication, which law applies?

The question is easiest to answer in respect of transactions to which the parties are all located within England. In this case it is likely that English law of evidence applies (see 8.2.1 as regards the place of formation of contracts where messages may have travelled through different countries en route. The relevant law of evidence is the law of the country in which the court case is being brought. With both parties located in England, any court case is likely to take place in England. If parties are located in different jurisdictions, then a court case could potentially take place in either jurisdiction or, possibly, in some third country.

8   Doing business on the Internet

8.1   Introduction

With a huge potential audience using the Internet, and the Web in particular, selling products and services over the Net can provide many benefits for organisations in fields such as advertising and service delivery. There is potential to reduce costs by cutting out the middle man and to provide a much quicker response to customer requests with fewer resources. However, there are also indications that business on the Net is not "taking off" to the extent that has been predicted.

Despite the hype surrounding the commercial possibilities of the Internet, buying and selling on the Internet is simply an extension of the traditional formation of a contract for the sale and purchase of supplies and services. Just as the law has evolved in the past to cope with technological advances in doing business, such as the telex and fax, so it will develop to meet the specific problems which arise from doing business on the Internet.

The concept of legally binding agreements, which can be enforced in the courts, is well established in all developed legal systems which have a set of rules built up through experience of adjudicating disputes which deal with the requirements to make an agreement legally binding.

In this chapter only the basic common law rules are addressed, and even they may vary somewhat between different common law jurisdictions. The principles outlined are as they apply to the common law of England and Wales. However, all common law jurisdictions have in common the requirement that there must be a "bargain"; namely, there must be offer, an acceptance and some form of consideration for the agreement.

8.2 Forming an on-line contract

8.2.1 Contract formation

There are three basic essentials to the creation of a legally binding contract:

- offer
- acceptance, and
- consideration.

This means there must be a sufficient degree of certainty to enable a court to ascertain the intentions of the parties concerned.

In broad terms, unless particular formalities such as writing are specifically required, a contract is formed when one party offers to do or to supply something on terms which are accepted finally and unequivocally by the other party, and that acceptance is communicated to the person making the offer. Something of value in the eyes of the law must be given to the person making the offer, and this generally takes the form of payment by the buyer for the supply of products or services. An intention to create a legally binding contract is usually obvious from the context. Whether there is sufficient certainty depends on the language used and the surrounding circumstances in each particular case.

Who may make a contract? Parties entering into a contract on the Internet may never meet or even talk to each other. In some circumstances, it may be of importance to the seller to know the identity of the buyer to ensure that a valid contract can be entered into. It may be that buyers need to provide by post, written evidence that they satisfy certain criteria before being issued with a password allowing them to place orders over the Internet. For example, the controller of an "adult" bulletin board or Web site may require proof of age.

Individuals Subject to certain exceptions, the general rule is that individuals may make contracts on their own behalf. The exceptions are as follows:

(a) Persons not of full age

The other party is bound but the minor is not bound unless he or she ratifies on attaining majority. However, contracts for the supply of necessary products (for example, clothing) or services (for example, education) are valid and may not be disclaimed when the minor attains majority.

(b) Persons not of full capacity

A contract with a mental patient is valid unless the other party knew that the disability prevented the patient from understanding the transaction, in which case the contract is voidable at the patient's option. If the patient's property is subject to the control of the court, the contract does not bind the patient but does bind the other party.

Contracts made with a drunkard are void if the drunkard is unable to understand the transaction due to extreme drunkenness and the other party knows this. However, drunkenness blurring business sense does not suffice to make the contract void.

Companies

In general, companies have the same contractual capacity as a person of full age and capacity. However, a company can only act within its corporate powers. Accordingly, it is necessary to ensure that the contractual obligations are not *ultra vires* and hence not binding upon the company. The company cannot avoid liability by claiming that the obligations were *ultra vires*. There are presumptions in favour of a person dealing with a company in good faith. For example, Section 9(1) of the European Communities Act 1972 provides that any transactions entered into by a director of a company are deemed to be transactions within the power of the company.

Third Parties

Products or services may be advertised and made available for purchase on the Internet via a third party "electronic retailer" such as CompuServe and Supernet. In the UK, Barclays group has launched BarclaySquare, a virtual electronic shopping mall with sites rented to a wide range of retailers. Under such arrangements, it may not always be clear whether a contract is formed with the supplier or the third party. The third party may be acting as an agent for a supplier or as a reseller. It is also possible that, in the event of a dispute and contrary to the intentions of the supplier and third party, a court may decide that the way in which the system operated meant that the third party was the contracting party rather than the supplier. Organisations need to be aware of this issue.

What are the terms of the contract

The starting point for the terms of the contract are what the parties have agreed to. In England and Wales, where the concept of the freedom of contract is recognised, the parties are free to agree between themselves whatever they choose, subject to a limited number of exceptions. For example, under the Unfair Contract Terms Act 1977 the law restricts the ability of a seller of goods to exclude responsibility for the condition of the goods and generally circumscribes the rights of both parties to limit or exclude their liability for breach of contract. Similarly, the Consumer Credit Act 1974 implies special conditions where one party is a consumer, particularly when credit is being given to that consumer by the seller of products and services.

However, other countries with civil law jurisdictions have based their legal system on the Roman legal philosophy of comprehensive codes. In Germany, for example, the commercial law revolves around the Civil Code and the Commercial Code. Here, the law implies many more provisions into a contract between the parties. It is worth noting that with the advent of EU law many more provisions are being implied in the UK; for example, the Unfair Terms in Consumer Contracts Directive and the UK Regulations implementing it.

In order for sellers to incorporate their terms within a contract, such terms must be brought to the attention of the buyer before the contract is formed. The more onerous or unusual the terms, the more clearly they have to be flagged to the buyer.

The conventional method of printing terms of business on the back of an order form or invoice is clearly inappropriate in relation to a seller's Web page on the Internet, and needs to be adapted. One method which has been used by companies is to refer prospective buyers to their legal terms by making use of hypertext. It is unclear whether this approach is sufficient given that many Web browsers probably do not click on this link before placing an order. Inclusion of specific terms and/or disclaimers at the head of a Web page is recommended (see 6.2 of the *Guideline* for advice on disclaimers). However, there is an uneasy tension between:

- what the law requires to notify the buyer of the seller's rights, and

- the marketing imperative to produce an interesting and fun page to capture the attention of the casual browser.

When is the contract made?

The exact point at which an exchange of electronic messages becomes a binding contract may be of considerable importance, not least in determining what the terms of the contract are. It may be that the seller wishes an order from a buyer to constitute the offer which the seller can accept or reject. This could be the case if there is a danger of stock of the relevant goods running out and the seller does not want to contract to provide something it may be unable to deliver. The seller may also want the right to reject orders from known bad payers or from certain jurisdictions.

In such circumstances, the Web page designer needs to make sure that the Web page is merely an advertisement, and that a potential buyer's communication is an offer, which the seller can either accept or reject. On the other hand, the seller may be comfortable that the statements on its Web page are the offer and that the order from the buyer is the acceptance which brings the contract into being.

The general rule is that a contract is not formed until the acceptance is communicated by the accepting party to the offering party. Where the communication is instantaneous, either face to face or over the telephone, the acceptance is deemed to be made at the time the message is received. This is often known as the "reception rule". In the example of communication by telex this is the case, even though the offering party may not read the telex until some time after it was received.

Conversely, where a contract is made by non-instantaneous means, such as post, acceptance is deemed to be effective at the moment the communication is sent. This is the so-called "postal rule". Thus, if there were problems with the postal service, the contract comes into existence even though the letter sent by the accepting party never reaches the offering party. However, this rule does not apply in most civil law countries.

There is no clear guidance as to when a contract is formed on the Internet. Where e-mail is used, is it possible that the "postal rule" applies to such an electronic message of acceptance? There are two justifications suggested for the "postal rule". The first is that it is a rule of convenience which has been devised for solving a difficult question. As discussed by Lord Brandon in *Brinkibon Ltd v Stahag Stahl und Stahlwarenhandelgesellschaft mbH* [1982] 1 All ER 293, this suggests that the "postal rule" might apply to electronic acceptances:

"The cases on acceptance by letter and telegram constitute an exception to the general principle of the law of contract [on grounds of expediency] .... That reason of commercial expediency applies to cases where there is bound to be a substantial interval between the time when the acceptance is sent and the time when it is received. In such cases the exception to the general rule is more convenient, and makes on the whole for greater fairness, than the rule itself would do".

The second justification is that the offering party has implicitly agreed that the accepting party may entrust the transmission of his or her acceptance to an independent third party, the postal authorities, and that therefore the accepting party has done all that the offering party requires for acceptance when the letter is posted.

This would suggest that acceptance takes place when the message is received by the service provider's computer rather than when the message is accessed by the recipient. This is further supported by the fact that there may be delay in delivery of the message on the Internet depending on the routing of the message and whether there is network congestion.

The nearest analogy to the Internet is with acceptance by telegram; it is necessary for the message actually to be communicated to the telegram service, normally by telephone (an instantaneous method of communication), but once it has been received by the service, acceptance is complete.

To try and reduce uncertainty on when a contract is formed, parties who trade electronically on a regular basis may decide to set up a special agreement (see 8.5.1).

Where is the contract made?

The place a contract is formed is mainly of interest in international transactions where the parties have not agreed which jurisdiction governs or, where there is no applicable international convention, to determine the jurisdiction. Again it is a question of whether the instantaneous or the non-instantaneous rules apply: in the former the acceptance takes effect at the place where it is received, in the latter the contract is made where the acceptance is sent.

This principle is further complicated on the Internet in cases involving corporate transactions. In such cases the computer of the offering company which first receives a message may be located in one jurisdiction, but that message is then automatically transmitted to a named individual in that company located in a different jurisdiction. Is the contract formed where the acceptance is first "communicated", even though it is unread and immediately re-transmitted, or where the human understanding of the acceptance takes place (the meeting of the minds required for contractual commitment)?

A number of international conventions determine, in respect of certain countries, the way in which their local laws apply to contracts involving their nationals; this is dealt with in Chapter 10.

8.2.2 Formalities

Generally, contracts do not need formalities in order to come into being. However, issues can arise concerning the evidence of agreement. The existence of terms helps resolve such issues.

Certain types of contract in various situations require particular formalities to be observed if they are to be enforceable. The most common of these are that the contract must be made or evidenced in writing or in a document, and that it must be signed. For example, any contract for the transfer of an interest in land. Note, however that the general rule is to the contrary; that is, oral agreements are binding.

Unless there is legislation which specifically provides to the contrary, "writing" under English law requires the communication to be in some visible form pursuant to the Interpretation Act 1978. This concept of a visible form casts a doubt as to whether words stored in an electronic form are "writing" within the meaning of the Act. However, if all that is required is a "document" then, unless this is also defined in the legislation or case law governing the transaction to require visible form, there seems no reason why it might not be produced electronically.

The requirement that certain documents be sealed has been almost completely removed by the Companies Act 1989 and the Law of Property (Miscellaneous Provisions) Act 1989. Seals, for example, were previously required to give effect to deeds. Obviously, a requirement that a seal be attached cannot be fulfilled in an electronic transaction. While the new provisions remove most of the requirements that the company's seal must be attached to make a transaction by that company effective, they do require signatures by two directors or a director and the company secretary.

A requirement for a signature is more problematic. Given that in may cases English law permits signatures to be typewritten or made via a stamp, there seems no reason to insist on a handwritten signature. Attention should instead be focused on the purpose of the signature; namely to authenticate the message as originating from the purported sender. If this is correct (and it must be recognised that there are no clear authorities on the matter), encryption offers the possibility of producing digital signatures that are more difficult to forge than handwriting.

8.2.3 Property rights

Certain types of documents are evidence of ownership of property, such as title to unregistered land. In some cases the mere fact of possession of certain documents carries the right to possession of property or payment; for example, bills of lading or bank drafts. It is clearly important to be able to identify the originals of such documents and to show that they have not been tampered with; both very difficult requirements to fulfil in an electronic context.

The Comité Maritime Internationale issued Uniform Rules in June 1990 for Electronic Bills of Lading and for Sea Waybills. These rules provide for electronic transmission of the data normally contained in such documents, with authentication provided by use of a private key issued by the carrier. In 1992 the Carriage of Goods By Sea Act was passed in the UK, which gives the Secretary of State powers to make regulations to provide for conventional documents used in this field to be in electronic form.

A possible way of dealing with proof of title is to replace the document with a central registration scheme (as many countries have done for title to land).

8.2.4   Record keeping

There are a number of statutory requirements relating to keeping business records. For example, a company is required to have records of its accounts and to keep them for at least six years (three years for private companies). VAT and tax records must be kept for at least six years. There are also common sense requirements; for example, where there is any possibility of a dispute, the records should be kept at least until the expiry of the relevant limitation period (6 years for ordinary contracts, and 12 years for deeds). Evidence of title to property must be retained as long as that property is owned, which could be a very long time.

While there are specific provisions for holding VAT records electronically, generally the statutory requirements do not specify the form in which the record is to be stored. One of the problems with storage of electronic data is that most of the media used for storage have not been around long enough to know how long such media can reliably retain data without corruption, or what the best storage conditions are. It is also important to be able to demonstrate that records have not been corrupted or tampered with.

8.3   Methods of payment

The speed of the Internet and its global nature are placing huge strains on the more traditional methods of payment. With recent estimates that credit card payments alone over the Web were worth $30 billion in 1995, there is clearly a need for a more appropriate method for moving around money. Most of the electronic payment systems have evolved in the US and, whilst they offer security, they do not always export well. For example, export versions of software products do not contain the same strength of embedded encryption as that available in the US. The methods of payment potentially available to buyers, which are described in the following sub-sections, can give rise to varying legal and security issues.

8.3.1   Traditional methods of payment

The buyer is requested via the Internet to send the seller cash or a cheque or other form of money order. This is likely to prove cumbersome, slow and costly, and many countries have exchange control regulations which prohibit the export of money. A variation is to provide credit card details via a telephone call using the public telephone system which is more secure than Internet telephony.

8.3.2   Established methods of payment over the Internet

The buyer is asked to e-mail credit card details to the seller, who has an agreement with the card issuer to accept payment in this way. The card issuer pays the seller and bills the buyer as usual. A debit card system would work in a similar way with the money being moved from one account to another by the card issuer.

The associated risk of this is that, unless the credit card details are sent in encrypted form, they can in theory be read by the operator of each server carrying the message and therefore there is scope for fraud. This risk already exists when payments are made using credit/debit cards over the telephone, but it is far less likely to happen over the public telephone system than over the less secure Internet. There is even a suggestion that some hackers have written software that monitors Web traffic and detects and records credit card transactions. The large credit card companies are currently investing heavily in encryption and other security techniques to avoid fraudulent use of cards.

From the point of view of the seller, there is also the question of authentication. How does the seller know that the person from whom the credit card details are received is the true owner of the card? One answer that a number of on-line retailers use is to form a club so that each shopper becomes a member with their own unique password. The member's card and other details need only be taken once and each time a purchase is made the password is used to verify and confirm the transaction. Of course, the password needs to be encrypted over the network to avoid the same fraud problem with the cards referred to above. Alternatively, public-key cryptography (pkc) can be used to provide a digital signature that can be authenticated through a trusted third party.

In March 1996 VISA and Mastercard launched Secure Electronic Transaction (SET) software which will ensure that both participants in a transaction are who they claim to be. When the buyer enters the credit card details into the computer, the software scrambles the numbers into a code which is transmitted to the seller.

8.3.3   Electronic Banks

At the end of 1994 First Virtual Holdings, advertised ambitiously as the world's first truly electronic bank, opened its (virtual) doors. The rudimentary transaction scheme run by First Virtual revolves around both buyer and seller holding their accounts with them. In the case of buyers, this amounts to an authority for First Virtual to make charges against their credit cards. When buyers, having investigated the seller's wares over the Internet and found something that they like, make a purchase, they give their account number to the seller, who ships the product.

Each day or each week the seller sends his list of who-bought-what to First Virtual, which sends e-mails to buyers asking them to confirm the transactions. Once buyers confirm, their (conventional) credit cards are charged and the money is transferred to the appropriate seller's account. If the buyer withholds confirmation, First Virtual withholds settlement.

This is really a sort of cut-out mechanism for people who want to shop on the Internet with their credit cards in a way that protects them against fraud. Because a buyer must confirm a transaction, much turns on trust; sellers may ship goods for which they do not get paid. Consequently its use currently only extends to low value items for which payment may not be made if the scheme did not exist.

8.3.4   Electronic cash

Here, electronic money-tokens are used similar to the kind packed into the pre-paid smartcards used for telephone calls. With the development of smart cards, credit card-sized devices containing a microprocessor can be filled and emptied of their store of digital money. The largest smart card trial is the Mondex project which began in Summer of 1995 in Swindon. Mondex cardholders use the smart card just like cash. They can use it for small purchases, such as a can of drink and larger purchases up to £500. The cards can be recharged with "cash" by inserting them into specially adapted automated teller machines or by using a Mondex telephone which transfers money from a bank account to the user's card. Mondex also allows users to transfer money from one card to another.

While the Mondex trial shows that smart cards and electronic cash are practical ways to pay for goods, the future of electronic money is moving from smart cards to a form that can be used easily over computer networks such as the Internet without the need for special cards or readers.

A company called Digicash has pioneered a system whereby the buyer converts real currency into electronic currency known as "cyberbucks" or "e-Cash" which only exists on the Internet. This is effectively a new currency which transcends national boundaries and can be used for transactions over the Internet and turned back into real money by the recipient. However, at present the currency itself is not backed by any national bank, in the way that every other currency is, and is still in a trial mode.

For true electronic cash to embody all of the properties that cash currently has, it needs to be verifiable as genuine and anonymous. This requirement can be achieved through the use of powerful encryption technologies in which digital signatures can authorise and verify payments whilst preserving the anonymity of the sender. However, anonymity and electronic distribution also raise other issues such as money laundering and evasion of taxes which could be made easier with electronic cash.

| 8.4 | **Charging for access** | Systems need to be put in place if content providers wish to be paid for content which is made available electronically. |

For example, the publisher of a magazine gains revenue through advertising space sold and through payment from buyers of the magazine. If the same magazine is made available on-line, potential buyers may access the magazine electronically rather than buying the hard copy. Of course, the level of advertising revenue may mean that the publisher can afford to allow free electronic access, although as take-up of the Internet increases, publishers are less likely to do so.

There may be a requirement to charge directly users who access information. A subscription mechanism facilitates such charging. For example, a Web site may merely contain a "snippet" or taste of an on-line magazine or study. The user is required to formally subscribe, by e-mail for example, and in return is given the key to unlock the information.

Members of organisations therefore need to prepare and present justification for their subscription to certain information.

8.5   Recommendations

8.5.1   Need for business
        protocols

Given the large number of legal issues which have not yet been satisfactorily addressed in relation to their implications for the Internet, as far as possible parties should enter into agreements setting out their "terms of doing business on the Internet" before they form a contract. After weighing up the associated legal risks, it may be the case that in terms of low value, low complexity or non-strategic products and services few terms are required, but for high value, high complexity or strategic products and services it may be preferable to make a contract in the conventional manner after which transactions pursuant to the contract can be carried out on the Internet.

Another possibility is that industry protocols can be developed for trading on the Internet in a similar manner to those relating to use of EDI. For example, in Europe industry groupings, such as ODETTE and CEFIC, have developed protocols for EDI. This has focused attention on the communications aspect of EDI rather than the underlying transaction. In relation to EDI the convention is to enter into an interchange agreement which binds the parties to a particular, structured form of communication.

Because different industry sectors inevitably have different specific requirements, no universal standard has been achievable. However, a number of organisations have produced model interchange agreements which provide a useful starting point for negotiations. On an international level the International Chamber of Commerce has produced the UNICID (**UNI**form Rules of **C**onduct for **I**nterchange of Trade **D**ata by Teletransmission) Rules. It should be noted, however, that such agreements may not bind third parties, or the courts, in the event of a dispute.

8.5.2 Allocation of risks/ liability for errors

It should be borne in mind that, unlike more traditional EDI, there may not be a network service provider substantial enough to take responsibility for secure and timely transfer of contract data. Although there is insufficient space here for a detailed examination of the form of the contract, it may be useful to set out the main areas which such an agreement should cover:

- the particular protocol for message formats

- the method for acknowledging messages and any confirmations of their content that are required

- which of the parties takes responsibility for the completeness and accuracy of the communications

- how payment is to be made and when it is deemed to be effective. (If a national currency is used, the seller should check whether there are any currency control regulations which would prevent the sale of supplies or services into certain jurisdictions. The contract should also address questions of exchange rate risk, although in most instantaneous transactions the seller simply states that payment must be made in its local currency and it is up to the buyer to find a means of paying in that currency)

- the security system to be used, and whose responsibility it is

- aspects of confidentiality

- data logs and the storage of messages

- authentication of identity

- which country's jurisdiction is to apply to the communications process

- dispute resolution; for example, agreeing to arbitration.

8.5.3 Payment validation

In a paper-based transaction system, the standard procedures for authorising payments include matching vendor invoices with the associated purchase orders and receiving documents. This lets the buyer verify that the goods were actually ordered, that they were received, and that the invoice includes charges for only those goods. The person who makes this check confirms it by signing the documents. On the Internet there may be no signature, and the confirmation document may not exist.

When developing these validation procedures, care must be taken to ensure that fraudulent or inaccurate transactions cannot slip through the validation process.

Identification and authorisation codes can replace signatures if the programmed control routine verifies that the electronic documents match. A code can be attached to indicate that the appropriate procedures have been successfully completed. With that step, the system has created evidence that the payment has been authorised.

8.5.4 Payment control

Paper-based transaction systems require signatures to authorise orders and payments. On the Internet there may not be anything presented for signature. That can increase the opportunities for unauthorised access. Anyone who gains access to the system can place orders or authorise payments to themselves or accomplices.

Vindictive employees or inadequately trained users can do things which have not been anticipated. In paper-based systems, the informal checking performed by people has served as a modest check on these contingencies. In an electronic system, another strategy is needed. The following procedures can help:

- creation of a file to hold purchase orders that require managerial approval. (There might be a policy, for example, that requires specific approval of unusually large orders or orders in excess of buyers' credit limits)

- use of multiple digital signatures when more than one signature is required to control payments

- encryption of files to prevent pirating of data or passwords

- development of computerised checks to simulate fraudulent activity.

9   Security

9.1   Introduction

"When making the decision to use the Internet for transferring files or for providing open access to other divisions, security is probably the most important issue that needs to be addressed. Some technologies such as EDI and X.400 incorporate a high intrinsic degree of security, but in the case of the Internet the responsibility for security currently lies with the user"[28].

The Internet has yet to become available to large sectors of the population. This in turn means that those who are users are often unaware of the possible security concerns. In many cases, organisations are attempting to impose security standards upon existing Internet users.

Control of the degree of access given to both internal employees and external third parties is an important issue for public sector organisations planning to connect to the Internet, particularly if the user is in charge and may have little interest in the security of the system.

There is no doubt that security is a problem on the Internet. The Internet is a massive source of information and gloriously wide-reaching but is also a source of uncertainty and potential disaster if staff are afforded unlimited access.

Computer security encompasses a number of areas which in turn raise legal issues, including:

- communications security

- contingency planning

- control of unauthorised access

- data control techniques

- data security

- encryption

- fraud/computer misuse

---

[28] "Public Sector Procurement and Finance" - November 1995 , issue 5 , p.16

- identification

- privacy and data protection

- risk analysis and management

- software protection

- viruses.

The above list is by no means exhaustive, and it is clear that laws on security in relation to IT and the information society have an increasingly important role to play in the workings of any organisation.

These include:

- the Data Protection Act 1984 and the proposed EC Data Protection Directive (see 3.9 and 3.8 respectively)

- the Computer Misuse Act 1990

- common law principles as applied to communications, privacy and confidentiality.

The Data Protection Act 1984 is in itself a relevant weapon in the fight against computer misuse; the Eighth Principle of the Act requires data processors to look after and maintain the confidentiality and integrity of the data which is processed (see 3.3.9).

It is important to appreciate that e-mails on the Internet are less secure than mobile phones or facsimiles. The Internet is not appropriate, therefore, for protectively marked correspondence or documents which may require security procedures to ensure authentication, integrity and confidentiality. Current Government guidance produced by CCTA[29] suggests that to use e-mail would be the equivalent of sending official mail on postcards.

---

[29] "Making the best use of the Internet" - CCTA April 1995, p.7

Another analogy is that sending e-mail is like risking an envelope being steamed open or stolen because it requires someone at an intermediate server quite deliberately to take steps to look at the mail passing through. E-mail is not intended to be public, other than to the recipient of the message, however, it should be remembered that unauthorised reading of e-mail is quite simple. E-mail is not the only manifestation of Internet security issues, as this chapter demonstrates.

9.2 Computer misuse

9.2.1 General hacking and security in general

IT security can be summarised in three concepts[30] which should always be borne in mind:

- confidentiality - prevention of the unauthorised disclosure of information

- integrity - prevention of the unauthorised amendment or deletion of information; and

- availability - prevention of unauthorised withholding of information or resources.

Hacking refers to the intentional unauthorised access, actual or attempted, to computer programs or data[31]. Comprehensive measures are needed to prevent and detect unauthorised access or disclosure (see 9.2.3).

Businesses are only just starting to become aware of the opportunities offered by the Internet in general and the Web in particular. Concerns about security are one of the things which are causing companies to be cautious, especially where financial transactions are concerned, but the technology is evolving to address those concerns.

---

[30] These are based on the DTI's security evaluation criteria first issued in 1989, although based upon the NCC guidelines published in 1979

[31] NAO Report on IT Security in Government Departments, 1995

As the NCC suggests in a recent guideline[32]:

"All you need to do is ask yourself whether the system is, or can be made, secure enough for your purposes. If not, leave well alone".

However, in government organisations the minimum standards of security and central security policy need to be considered.

Notes from the US General Accounting Office Report to the Congress[33], published in January 1995 identify "insiders" as a main source of security violations, either in terms of accessing information within the organisation, or disclosing information to outsiders. The report states:

"Many violations of information safeguards are perpetrated by trusted personnel who engage in unauthorised activities or activities that exceed their authority. These insiders may copy, steal or sabotage information, yet their actions may remain undetected."

Given that security problems are likely to come from within, all organisations must evaluate what they perceive to be the benefits to be gained from the Internet and balance them with their obligations to their profession and/or their obligations under general law.

Whilst hacking has existed for as long as electronic data communications have existed, in the UK, it was not until 1990 that specific legislation was passed to deal with the problem. As a result of a 1989 Law Commission Report, the Computer Misuse Act joined the statute book in 1990.

The Act makes it a criminal offence for anyone to access or modify computer programs or data, or to attempt to do so, without the authority of the owner. It provides a framework for the prosecution of people who misuse computers. It should be borne in mind, however, that the

---

[32] NCC Guideline No. 192, "Internet Security - an introduction to the principles and practice", May 1994

[33] "Information Superhighway - An Overview of the Technology Challenges", p.18

Act can only complement existing security measures, not replace them. Security measures should clearly state who is authorised to use a system and the degree of that authorisation.

9.2.2 Offences under the Computer Misuse Act 1990

Unauthorised access offence

This offence was created to deal with the problem of hacking, but it also deals with authorised users who access parts of the computer system to which they are not authorised. For example, a person may be authorised to access personnel data files but not financial ones. If a financial file is accessed without authority then an offence has been committed.

Section 1(1) of the Act states that:

"A person is guilty of an offence if:

(a) he causes any computer to perform any function with intent to secure access to any program or data held in a computer;

(b) the access he intends to secure is unauthorised; and

(c) he knows at the time that he causes the function that that is the case".

To prove an offence has been committed it is necessary to show that the :

- access was deliberate

- access was unauthorised

- person carrying out the offence knew that it was unauthorised.

Many possible excuses are removed by Section 1(2) which states that:

"The intent a person has to have to commit an offence under this Section need not be directed at:

(a) any particular program or data;

(b) a program or data of any particular kind; or

(c) a program or data held in any particular computer."

A person found guilty of an offence under Section 1 would be liable to a fine of up to £2000 or imprisonment for up to six months or both of these.

Ulterior intent offence

Unauthorised access for the purpose of committing a serious crime is viewed more gravely.

Section 2(1) of the Act states that:

"A person is guilty of an offence under this Section if he commits the "unauthorised access offence" with intent:

    (a) to commit an offence; or

    (b) to facilitate the commission of an offence (whether by himself or by some other person)."

To prove the ulterior intent offence it must be shown that the accused:

- deliberately accessed the computer

- did not have authority to do so

- knew that authorisation was not given to access the computer.

A person found guilty of an offence under Section 2 of the Act is liable to a maximum sentence of five years imprisonment and/or an unlimited fine.

Unauthorised modification offence

The unauthorised modification offence means causing any modification of programs or data held in the computer knowing that this is unauthorised and with the deliberate intent to impair the operation of the computer. The introduction of viruses, Trojan horses, logic and time bombs is covered by this section of the Act.

It is immaterial under the Act if the modification is permanent or temporary. The only point of importance is that there is an intent to impair the computer system and it does not matter if the damage is caused immediately or in the future.

This offence is covered by Section 3 of the Act, which states that:

"A person is guilty of an offence if:

    (a) he does any act which causes an unauthorised modification of the contents of any computer; and

(b) *at the time when he does the act he has the requisite intent and the requisite knowledge."*

To be proved guilty the modifier must have the "requisite knowledge" that he or she is unauthorised to carry out the change and the "requisite intent" must be malicious. This is further defined in Section 3(2) which states that:

"The requisite intent is an intent to cause a modification of the contents of any computer and by so doing:

(a) *to impair the operation of the computer;*

(b) *to prevent or hinder access to any program or data held in any computer; or*

(c) *to impair the operation of any such program or the reliability of any such data."*

Section 3(3) of the Act states that:

"The intent need not be directed at:

(a) *any particular computer;*

(b) *any particular program or data or a program or data of any particular kind; or*

(c) *any particular modification or modification of any particular kind."*

A person found guilty of an offence under Section 3 of the Act can be imprisoned for up to five years and/or be subject to an unlimited fine.

Proving the offence

To prove that unauthorised access has taken place it is necessary for the computer system to have an access control system with a secure log to record all significant events. The log can be used to prove that access was attempted, or successfully made, from a particular workstation at a certain time.

It is more difficult to prove that the access was made with ulterior intent. When the nature of the system or data accessed has been established, a check should be to see if any action such as the setting up of bank accounts has been made.

The proof of unauthorised modification is assisted if the time at which this took place can be established. Backups and printouts that are taken on a regular basis can be of assistance here. The access control system should record when data is accessed, when it is backed-up and when it is updated. All of those using the computer system should be aware of the limits set for them to make modifications to data and programs.

9.2.3 Measures to prevent computer misuse

There are a number of ways in which computer systems are vulnerable. Amongst these are the mistakes that personnel make, faults in hardware and software, and hackers who can access systems and either steal information or corrupt data files. There are viruses that are self-replicating pieces of computer code that act by corrupting programs or data files. Finally, there is the theft of hardware, software and manuals.

To prevent misuse, organisations must as a minimum have an IT security policy. In the case of government organisations this policy needs to encompass the following six main criteria:

- establish and enforce a departmental IT security policy consistent with central government security policy and standards

- maintain an organisation to direct and manage IT security

- ensure that risks are reduced to an acceptable level by applying protective measures which are based on risk assessment and the protective marking of information, and which conform to appropriate minimum standards

- limit access to information and other IT assets to those whose duties require it and who have the necessary authority and security clearance

- ensure that personnel are aware of IT security policy and practice to the extent that their duties require and fully understand their responsibilities (including their legal obligations), and

- monitor and review IT security arrangements to ensure that policy, standards and procedures remain relevant and effective.

In some situations it may be necessary to include in the security policy, reference to specific access control mechanisms which have been determined by risk analysis and management methods such as CRAMM.

Specific countermeasures which can be adopted are:

- restrict access to the computer system and the information stored within it

- use passwords which allow only authorised people to access the data files

- have an access control system that records any access, or attempted access, that is made, by whom and which files were accessed

- check the references of all employees before they are allowed to start work

- in systems that are susceptible to fraud, such as payment systems, do not allow any one person to have the authority to access all of the functions within the system

- have in-built checks within systems so that high value transactions, variation in volumes of transactions and invalid attempts to access the system are all reported

- protect access to the data files by the use of passwords

- ensure that passwords are made up of a mixture of numbers and letters, that they are not written down, and that they are changed frequently

- have a secondary password for additional protection for sensitive data

Risk analysis and management can also be used to determine more precisely appropriate countermeasures. For example, against unauthorised access, biometric based access control, which includes unique personal identification such as retina scan or thumbprint, may be appropriate to improve accountability and auditability.

9.2.4 Use of the Computer Misuse Act 1990

It seems clear that the business community is relatively unaware of the existence of the Act. A 1993 DTI report cited several reasons for the unwillingness of organisations to bring a case under the Act which included:

- fear that a prosecution might damage their reputation

- belief that the investigation and a subsequent prosecution would waste their time

- doubts over whether there would be any benefit from reporting the crime

- concerns about the police's technical ability to investigate cases of computer misuse[34].

9.3 Computer viruses

All IT systems are particularly vulnerable to attack from malicious software such as viruses which, once loaded, can cause serious damage to data and programs. The "openness" of the Internet means that there is greater opportunity for the spread of viruses and organisations need to be aware of the dangers.

9.4 Other legal requirements

In association with the Computer Misuse Act 1990 and Data Protection Act 1984, there is a whole area of law concerning confidentiality, liability and integrity which requires compliance. Such law is to be found in various business sectors.

For example, under the Financial Services Act 1986 Section 47A, the Securities Investment Board issued Statements of Principle on 15 March 1990. Rule 3-12 mandates that, organisations:

"must establish and maintain at all times effective systems of internal control."

Therefore, the requirements for security measures go beyond the requirements of common sense in the information society and enter the realms of mandatory rules.

---

[34] "IT Security - the Legal Challenge", John Worthy - Computer Law and Security Report - March-April 1995, p.65

9.5   Confidentiality

The duty of confidence derives from both contract law and the law of tort.

Under a contract, one party may designate certain important information as "confidential", and thereby, establish a contractual right to injunct the other party for a threatened disclosure or misuse of the confidential information or claim damages for an actual disclosure or misuse.

Tort is a breach of legal duty, other than under contract, and deals with civil wrongs; for example, negligence, liability for damages.

In tort, an individual or organisation may be under a duty to keep certain information confidential; perhaps as a result of guidelines written into codes of conduct. The party to whom the duty of confidence is owed may obtain an injunction to prevent a threatened disclosure or misuse of confidential information or claim damages for an actual disclosure or misuse.

A successful breach of confidence action requires the presence of three elements:

- the information must be of a confidential nature

- the recipient of the information must be aware that an obligation exists

- an unauthorised act must have occurred or be about to occur.

However, the international nature of the Internet can limit the capability of organisations to instigate an action for breach of confidence. Unlike other intellectual property rights, an action for breach cannot be brought against a person operating in another country. In addition, the publication and distribution of confidential information abroad can invalidate the enforceability of protection within the UK (see also 10.2 for more information on laws governing international transmissions).

The implications for Internet usage are legion. Consider the possible liability for release of valuable confidential information in a Usenet posting. Further, consider the liability which may attach to an organisation for a posting by its employee.

The law relating to confidence and an appreciation for the commercial value of certain information, needs to be taken into account by organisations wanting to regulate Internet usage by its members/employees.

9.6 Fiduciary duties

A fiduciary duty is a legal term to explain a special relationship which various industry sectors and professions deem to exist between their members and their clients; the general public. The duty is a higher duty than that owed by one citizen to another. It arises due to the fact that the member's profession assumes a status upon which the general public are entitled to rely.

Various industry sectors and professions feature such duties. The level of duty is not enshrined in statute, although certain areas of law, for example the tort of negligence, use the existence of such a duty to imply a higher duty of care than that normally expected of the general public to each other.

For example, a law firm has a duty of trust, a fiduciary duty to keep its clients' information confidential. Therefore a normal Internet connection, available to all staff on all terminals, allowing direct 2-way access is not allowable. At the same time communication with a client over the Internet may be tremendously advantageous, provided that the client is warned and accepts the risks of communication in this way.

Therefore, any fiduciary duty that may exist must be considered by any organisation contemplating Internet usage.

9.7 Liability for loss

Liability becomes an issue in any discussion of security because a breach of security may lead to loss, usually financial, and injured parties may look to an organisation to make good that loss.

9.7.1 Vicarious liability

In general terms, organisations may be held vicariously liable for the acts of its employees undertaken in the course of their employment. A classic example is in the tort of negligence, where a company may be liable for damages to a third party for negligent acts of its employees if the employees concerned were acting within the scope of their employment.

Vicarious liability is strict in the sense that the employer need not be guilty of personal fault. The imposition of vicarious liability is based on employers' ability to supervise the acts of their employees.

A wrongful act is an act done in the course of employment if it is either:

- a wrongful act authorised by the employer; or

- a wrongful and unauthorised mode of doing some act authorised by the employer.

In the Internet sense, the possibility is that employees may, via e-mails or Usenet groups, bind their employers to a contract or opinion, or make representations ostensibly on behalf of their organisation, or cause damage for which their employers can be held liable. This obviously needs good supervision.

9.7.2 Employee liability

It is not always the case that an injured party can look to the employer to make up for the negligence of an employee. Generally, employees must have caused negligence in the course of their employment. If employees did something frivolous or, perhaps, totally unconnected with their employment, then the employer is not liable.

9.7.3 Grey areas

There are grey areas between the above two extremes. For example, what is the situation where employees:

- e-mail colleagues in the course of their work but use slack and, perhaps, defamatory language?

- post information on behalf of their company which infringes a third party's copyright?

Whether the company is vicariously liable in such instances depends on the facts of the particular case. Therefore, it is apparent that controlling staff access to the Internet, and supervising how they use it, are real issues related to system security generally.

9.7.4 Causes of liability

Whether liability arises through negligence or misrepresentation, allocation of responsibility for errors or inaccuracies is an issue. Thus information which is included in, or supplied via, a network may be inaccurate in the first place (or may degrade with time). Also there is the perhaps greater risk for viewers or users to interact with the information on offer and to corrupt the information, leading to misleading particulars being disseminated to subsequent users. For example, who would wish to take responsibility for information posted freely on the Internet? That information could have been used, distorted and manipulated by different persons with differing levels of expertise and/or intentions.

9.7.5   Apportionment of loss

Where corruption (or loss) of information gives rise to financial loss or other harm, the question arises as to how that loss is to be apportioned between the various parties involved.

By way of example, Large Co. may offer via the Internet an on-line holiday booking service. This may include pictures of the various locations, together with facts and figures. These may include statistics on hours of sunshine and hotel rates. Large Co. must be able to deliver; that is to say it must not misrepresent any of the facts and thereby falsely induce a customer to enter into a contract to buy that holiday. If the holiday does not live up to the promises made then Large Co. may be liable to pay damages to the aggrieved customer.

Such a service provider may be able to limit liability by including appropriate disclaimers. Under general principles of contract law, such disclaimers must be clear and visible, and must take effect before the viewer/user enters into the contract. Business users are more likely to be treated as being bound to contract law than private individual users (see Chapter 8.2.1).

9.8   Guidelines and their status

Organisations need to consider what guidelines to issue to employees in relation to their use of the Internet.

Whatever guidelines are created, it is necessary to be sure of their status. If the guidelines are incorporated into employees' contracts of employment, this means that consideration needs to be given to which breaches of contract entitle an employer to dismiss an employee.

Employers should update staff handbooks and/or standard terms and conditions to incorporate statements as to what is considered proper use of computer systems on the organisation's behalf; for example:

- who is allowed to trade electronically in the company's name

- how should that trade should be conducted

- what limits or bans should there be on inappropriate use, such as private use of any kind or accessing illegal or obscene material

- what rules are necessary to try and prevent the introduction of viruses.

These variations to existing contractual terms should cause no problem legally. Such guidelines do not affect pay or status making it difficult for an employee to claim the variation of terms is serious enough to justify leaving and claiming constructive dismissal (generally, in the absence of agreement a variation of terms is not binding and an employee can treat the variation as constructive dismissal). The employee gets the benefit of the technology and so would find it hard to argue that the variation is detrimental.

Such guidelines could be given the status of "codes of practice" which would not be part of the employment contract and hence not strictly terms of the employees' employment, but would be relevant in determining whether an employee had acted properly when considering unfair dismissal.

9.9 Possible security measures

9.9.1 General

Organisations should seek to protect their sensitive information and security measures help to ensure privacy. Reducing the frequency and damage of attacks is the key and the following focal points should help:

- identification and authentication: knowing the user's identity and the message's authenticity

- access control and authorisation: protecting information from unauthorised access

- confidentiality: protecting information from unauthorised disclosure

- integrity: protecting information from unauthorised modification or accidental loss

- nonrepudiation: ability to prevent senders denying they have sent messages; and

- availability: ability to prevent denial of service; that is, to ensure that service to authorised users is not disrupted when it is required.

Internal mechanisms

Where the Internet is to be used by staff dealing remotely with the personal affairs of the public, mechanisms need to be set up to ensure that staff can verify that they are dealing with the correct person and not disclosing personal information to an imposter.

Making employees aware of security is essential. Organisations should try to run regular refresher sessions. In addition, the use of posters and stickers on notice boards and PCs can help.

At the same time as supervising employees, organisations must control their information. Classification of data into varying degrees of sensitivity can be matched with access rights so that, ideally, only a few senior personnel have access rights to highly sensitive information and only authorised personnel have the power to give opinions in newsgroups on behalf of their organisation.

The concepts of classification of data and access rights are well understood in Government. On the one hand, we have the Protective Marking System (PMS) and, on the other, we have security clearances and the need-to-know principle.

Commercial software

With these guidelines in mind, what can the marketplace offer? There is software currently available which allows limited access to undesirable newsgroups or sites. There are also secure network applications whereby access to newsgroups can be strictly controlled and which provide facilities to monitor incoming traffic (such as e-mails) to staff, students, and other potential receivers.

Security

9.9.2   Firewalls

Technology, as well as creating new legal problems, has also provided a number of solutions. For example, many companies employ "firewalls".

A firewall is a collection of hardware and software components that together provide a protective channel between networks with differing security policies. Legitimate communication may be made only through this protective channel, and when such communication takes place, it is tightly controlled and heavily audited. Attempts at unauthorised communication can be detected though not necessarily prevented. A firewall can allow restricted access to certain Internet port numbers and block access to everything else.

The security performance of commercially available firewalls has not been formally evaluated, but their effectiveness is known to vary greatly. Experts acknowledge that even the better products may not be fully effective in preventing unauthorised access and stopping malicious code being mailed into a system.

Despite these shortcomings, firewalls have been found useful in corporate and commercial applications, and have a role to play in selected lower risk Government situations. If the data to be protected against risks arising from direct connection to the Internet does not exceed "Restricted", a firewall or comparable device can be accepted.

Organisations considering following this path need to have a clear understanding of the risk factors, and how in practice the proposed policing device will contribute to controlling the risk. Government organisations are strongly advised to discuss their needs with the Government's security authorities and to ensure that any connections they establish to open services are made through a reputable service provider.

Some organisations may feel compelled to set up a standalone PC or network, separate from the main network; in other words to introduce an "air gap" between the Internet and the organisation's network. It is a matter of the degree of risk that the organisation wants to accept, always bearing in mind the obligations under the Eighth Principle of the Data Protection Act 1984 (see 3.3.9).

9.9.3   Encryption

Encryption can play a key role in ensuring a secure network. It is a method of transforming ordinary plain text into a form that only the intended recipient can decipher (this is commonly known as ciphertext).

Moves in the United States such as the "Clipper Chip Controversy" have sparked a great deal of concern over the rights of individuals to privacy as the US Government have demanded that their law enforcement agencies be allowed access to any key. Clipper is based on a secret cryptographic algorithm known only to the US Government, who also have access to the keys through US Government key escrow agencies. There is a similar initiative being considered at the European level but using public key escrow. Public-key cryptography (pkc) is a powerful technique for encrypting, hashing and signing messages or files, but it is only as good as its implementation.

By contrast, the proposed UK initiative is based on commercial, as opposed to "public", key escrow using a cryptographic algorithm chosen by the parties concerned; providing that this is also put into some form of escrow. The law enforcement agencies would have access to the keys under the authority of a Court Warrant. The idea is not to mandate the cryptographic algorithm and at the same time not weaken the ability of the law enforcement agencies to investigate crime.

CCTA has been working in conjunction with the Communications-Electronics Security Group within Government to devise a means of making information transmitted on the Internet secure. The initial solution for Government "Restricted" level information is a product called Secrets for Windows (HMG Version). This product is now available commercially.

The choice of algorithm for use within Government to protect official information is determined by the security authorities. Currently, the Digital Signature Algorithm (DSA) with its associated Secure Hashing Algorithm (SHA) is approved for hashing and signing messages and files, though the encryption algorithm depends upon the sensitivity of the information.

One example of encryption technology available on the Internet is Privacy Enhanced Mail (PEM), the Internet *de facto* standard for secure e-mail, which provides encryption as well as user authentication and is publicly available. The contents of individual Web pages can be encrypted using Secure Hyper-Text Transfer Protocol (S-HTTP) without affecting the identifying server information. In both examples, the encrypted information is able to pass through organisation firewalls.

The Council of Europe on 8 September 1995 recommended that measures should be considered to minimise the negative effects of the use of cryptography (for example, its use to cover up criminal activities) without affecting its legitimate use more than is strictly necessary. Such proposals could make telecommunications operators responsible for decrypting traffic and supplying it to Governments when asked. This could also mean changing national laws to enable judicial authorities to chase hackers across borders.

9.9.4 Digital signatures

Digital signatures are an encryption technique which enables a message to be tagged with a unique identifier, which can be recognised by the recipient. Generating a digital signature is referred to as "signing" the work and serves as a means for authenticating the work, both as to the identity of the person that "signed" it, and as to the contents of the file. The signature is computed on the work being protected, the signature algorithm and a personal key. It is unique for each different item of work for which it acts as a "seal".

It should be remembered, however, that the sender is responsible for the security of the digital signature. The recipient cannot know whether a signature has been deliberately or negligently disclosed. However, the recipient could be protected by entering into an appropriate contractual framework for using the system with the sender, incorporating appropriate safeguards for both the sender and recipient.

Digital writing in general The status in law of digital writing as opposed to hand writing on paper, is open to interpretation. However, it is clear that electronic evidence may be treated as evidence in court under certain circumstances. Similarly, a digital literary work still attracts literary copyright and a libel in electronic form can still be an actionable libel. Therefore, there seems to be little difference. "Writing" under the Interpretation Act 1978 is defined in terms that there is a requirement for something to be in visible form. By this analysis, writing in digital form (at least when called up on screen) is sufficiently visible to constitute writing for the purposes of making contracts.

Legal enforceability of digital signatures The IDA programme[35] commented recently:

"When studying the legal implications of electronic procedures in the member states, ... different meanings are given to the concept "signature", depending on the state of computerisation in the member states."

There is a link between the requirement for signature and the requirement for writing. Once a contract is in written form, the signature is the "sign-off"; the identifying mark which is aimed to state that a signed document is endorsed by the person signing it. However, increasingly signatures can be stamped or typed and there are many cases where there appears to be no need to insist upon physical handwriting. Therefore, rather than focusing on whether the signature has been produced by man or machine, it is necessary to focus instead upon authentication (that is, the purpose of the signature). The ideal way to prove the authenticity of an electronic signature is through the use of cryptography.

Using encryption techniques, particularly those that rely on the use of both public and private keys, can be an effective way to utilise electronic signatures in order to replace handwritten signatures without losing the basic requirements of a signature. However, organisations need to implement clear internal guidelines concerning

---

[35] "Legal Aspects of the Interchange of Data Between Administrations" - draft reports November 1995, p.11, copyright ICRI-K.U. Leuven

the security and use of digital signatures in a contractual situation and where there are particular evidential requirements.

Legitimate concerns

Organisations who transmit digitally signed documents may have legitimate concerns such as:

- are there any circumstances in which senders could reasonably disclaim responsibility for an unauthorised transmission which bears their signature?

- what can senders do to protect themselves after a breach of security has taken place and their signature is no longer secure?

With regard to the former, unless the recipient has actual knowledge of the misuse, the sender is likely to be bound by all transmissions which incorporate his or her digital signature. In the case of the latter, once a user of a digital signature has detected a breach of security, immediate and effective steps should be taken to prevent the misuse by notifying all potential recipients.

10 The international dimension

10.1 Introduction

The Internet brings into focus issues relating to the treatment of international flows of information under national systems of laws. These issues are, to a great extent, not novel; traditional publishing has from its inception, been an international industry and satellite broadcasting raises distinct issues in relation to national treatment of international transmissions.

In particular, the supra-national status of the Internet highlights issues relating to those countries' laws which apply to electronic transmissions, how variations in national laws apply to international transmissions, and which courts have jurisdiction over international transmissions.

In relation to government transmissions, there are specific issues relating to cross-border transmissions. In particular, there are issues relating to compliance with the legal framework for electronic transmissions and also the compliance with procedural requirements of the transmitter and receiver.

10.2 Law governing international transmissions

10.2.1 Some rules on cross-border transmissions

When Internet transmissions cross one or more national boundaries, which countries' laws govern those transmissions? In general terms, the answer to this question is that it depends on the law of countries in which the transmissions are posted, pass through and are accessed. Each particular situation must be analysed in order to determine the applicable law in each instance. However, there are a number of generally applicable rules:

- *applicability of the law of the country of posting of the transmission*

  Wherever a transmission is posted it is subject to the laws of that country. For example, if a transmission is made from a sender in the UK, to a recipient in France which unauthorisedly discloses personal data, UK data protection law clearly applies to the international transmission

- *applicability of the law of the country of access*

  In a great many instances the law of the country in which a transmission is accessed applies to that transmission. Taking the previous example of a transmission posted in the UK, but accessed in France, if that transmission includes unauthorised images of a public figure, the French privacy laws may well be infringed and the sender of the transmission may be subject to legal action in France

- *applicability of the law of the country or countries of transmission*

  Where electronic messages are only transmitted through countries (and not accessed) the transmissions are less likely to be subject to the laws of the countries of transmission. For example, where an e-mail transmission containing defamatory material is transmitted using "packet switching" technology through a number of countries before reaching the intended recipient, as the transmission remains in electronic form no defamation is actually made. However, as soon as the transmission is accessed, it is subject to the law of the country of access.

10.2.2 Varying legal requirements

An inevitable corollary of the national application of laws to the Internet is that different legal requirements apply in different countries. Since national law applies to transmissions, the varying legal requirements applying to Internet transmissions means that the same transmission can be treated in a variety of ways in different countries.

The following examples illustrate this issue:

- advertising which is legal in one country may be illegal in other countries. For example, advertising alcohol in the Middle East or Coca-Cola in Albania is illegal but perfectly acceptable in the UK. Many countries have advertising restrictions on otherwise legal products; for example, tobacco, financial services, and contraception

- intellectual property rights are subject to national treatment, which means that infringement is assessed in relation to the law where the infringement occurred. Hence, where there are differing treatments of IPR, a transmission may be an infringement in one country and not in another and vice-versa. For example, at present the copyright protection provided to databases in the UK is considerably stronger than that provided in Germany. Hence, a database may be developed in Germany which does not infringe German copyright legislation but, if electronically transmitted to the UK, might infringe the copyright of a UK database that attracted copyright protection

• pornography and obscenity laws are notorious for their national variations; material which is regarded as pornographic under the law in the UK may well be legally accessed in a number of Scandinavian countries.

These varying national legal requirements mean that there is, inevitably, a degree of uncertainty in relation to the compliance with the law when transmissions are made on the Internet. The technology infrastructure may mean that the sender of information has no control over countries from which the information can be accessed. For example, if a home page is established on the Internet, the home page can be accessed by any Internet user in any country. This has a potential risk that the laws of any country might apply to the contents of that home page.

Suggestions have been made that in order to overcome the difficulties posed by the national treatment of Internet transmissions, a single body of law should apply to the Internet. This would have the advantage that there would be no varying national treatment but would have the disadvantage that in particular countries electronic transmissions could be treated differently to hard copy publications. This could give rise to a situation where electronic transmissions would be legal, whereas hard-copy publications would be illegal.

In any event, the establishment of such a "cyber-law" could only be established by means of inter-governmental treaties. Since the international acceptance of copyright has taken over 100 years to become virtually universal, the establishment of "cyber-law" in a wide variety of areas is likely to take a considerable time. At present, there does not appear to be an economic motive for the establishment of such an international "cyber-law".

10.2.3 Court jurisdiction

In addition to the issue of the applicable law, there is also an issue relating to whether the courts of a particular country have jurisdiction over particular issues. As with the issue of the applicable law, in general terms the answer to the question is "which courts have jurisdiction?" must be analysed in each instance in order to determine whether a particular court has jurisdiction in relation to a particular situation. Again, there are a number of generally applicable rules:

- *jurisdiction of the court in the country of the posting of the transmission*

  In general, wherever a transmission is posted, the courts of that country will have jurisdiction in respect of the transmission. For example, if an issue arises relating to the disclosure of personal data in relation to a transmission originating in the UK and transmitted to France, the English, Scottish or Northern Irish courts would clearly have jurisdiction (as appropriate)

- *jurisdiction of the court in the country of access*

  In general, the courts of the country where a transmission is accessed have jurisdiction. For example, if a transmission accessed in France contravenes French privacy laws, the French courts would have jurisdiction

- *jurisdiction of the court of the country or countries of transmission*

  It is less likely that a court in a country through which a message has been transmitted would consider that it had jurisdiction. However, this question would need to be considered on a case-by-case basis.

More difficult issues may arise when the fact of varying legal requirements becomes a real issue. For example, if a transmission originating in the UK contravenes French privacy law, but the sender is a private individual and does not have a presence in France, there could be difficulties bringing a court action in France against the sender who is not located within that jurisdiction. In these circumstances, it would be possible to bring an action in the English courts against the sender. However, the English courts would not apply French privacy law to a transmission which originated in the UK, but which was accessed in France.

In these circumstances, if an action could be brought against the sender in France and the action was upheld, the judgment of a French court would be enforceable in the UK. Court orders for the enforcement of French judgments in the UK are obtainable from the English courts because both France and the UK are signatories to the Brussels and the Lugarno Conventions relating to the enforceability of the judgments of overseas courts.

In relation to the enforceability of overseas court orders, one significant anomaly at present is that the US is not a signatory to the Brussels Convention. Hence, if the judgment of an English court needs to be enforced in the US, the US court is entitled to re-examine the issues. If the judgment is based on a law which could not be applicable in the US, the US court is unlikely to order the enforcement of the overseas judgment.

10.2.4 Export control legislation	Certain types of information may be regulated under export control legislation; for example, Export of Goods (Control)(Amendment) Order 1995. The export of this regulated information requires an export licence from the Department of Trade and Industry.

The use of cryptographic techniques to authenticate or to provide for confidentiality for electronic messages sent internationally may also be restricted by export regulations, both within the UK and other jurisdictions; for example, US and France.

10.3 Legal framework for governmental transmissions	In a number of continental jurisdictions it appears that, in some instances, there must be a positive legal authorisation in order for valid electronic transmissions to be made by a civil servant in respect of government business. Where these requirements apply, if a UK civil servant transmitted an electronic message to a counterpart in one of these countries in a manner which

did not comply with the requirements of the legal framework, then there is a risk that such transmissions would not have been valid.

It is likely that in the ordinary course of events such technical non-compliance would not lead to significant difficulties. However, if such transmissions became significant in relation to a legal action it would seem that any technical non-compliance could affect the validity of such messages.

This issue was investigated under the IDA programme (Legal Aspects of the Interchange of Data between Administration). A draft report by the University of Leuven was published in November 1995 together with draft guidelines. In broad terms, the guidelines recommend that cross-border administrative co-operation be exhibited by the signing of interchange agreements. Such agreements are intended to resolve the anomalies between member states' public authorities. The guidelines suggest that the agreements should be based upon the European EDI Model Agreement 1994[36].

10.4 Administrative procedural requirements

A number of continental jurisdictions have stricter requirements than the UK in relation to the procedures by which administrative decisions are made. For example, there may be requirements that administrative decisions are sent to the members of public concerned on paper, and that the decision is signed by the appropriate civil servant.

There may be issues in relation to cross-border electronic transmissions if these transmissions mean that the administrative procedures in the recipient state are not complied with. The Leuven guidelines suggest a marriage between technical and legal requirements in order to preserve the legal validity of communications by appropriate security mechanisms (see Chapter 9 on security generally). It is to be hoped that Interchange agreements will remove the differences between EU member states.

---

[36] The revision of the final draft of a "European Model EDI Agreement" open to comment in May 1991 has been adopted as Commission Recommendation of 19 October 1994 relating to the legal aspects of electronic data interchange, Official Journal (1994) L338/98

Glossary

Air gap	Where an internal network is kept physically separate from the Internet.
ASA	Advertising Standards Authority.
BABT	British Approvals Board for Telecommunications.
Bulletin board	A repository for messages and files, often devoted to a particular topic. Bulletin board systems can provide wider facilities including a forum for discussion.
CCTA	Central Computer and Telecommunications Agency.
CDPA	Copyright, Designs and Patents Act 1988.
CEFIC	A project in the chemical industry sector set up by the European Council of Chemical Manufacturers' Federation.
CIMTECH	The Centre for Information Management & Technology - University of Hertfordshire.
CRAMM	CCTA Risk Analysis and Management Method. A complete method for identifying and justifying all the necessary protective measures to ensure the security of IT systems.
Cyber cafe	A place where facilities are available for customers to access the Internet as well as having some refreshment.
cyberspace	A term used by William Gibson in his novel, Neuromancer. It is often used to describe an electronic universe of information on the Internet or the associated culture.
DEO letter	Dear Establishment Officer letter.
EC	European Commission.
EDI	Electronic Data Interchange. The transfer, from computer to computer, of data using an agreed standard to structure the message content.
E-mail	Messages sent by a user over networks to other users.
EU	European Union.

Ex parte	The situation where an application is made to court and an order received purely by one party to an action; that is, the party against whom the order is sought does not need to be told of the application. An *ex parte* order can therefore be beneficial where the element of surprise is required.
FTP	File transfer protocol. A protocol that defines how to transfer files from one computer to another. It can also mean the program which serves the files using the protocol.
Home page	Usually the first page of an Internet site seen by a reader (browser), which serves as an introduction to the organisation.
HMSO	Her Majesty's Stationery Office.
HTML	HyperText Mark-up Language. The language in which World Wide Web documents, often referred to as pages, are written and which includes computer instructions on displaying the information on the pages.
HTTP	HyperText Transfer Protocol. The network protocol used for transferring data on the World Wide Web.
Hypertext	Text that contains electronic links to other text, files, pictures etc. within the same document or other documents. Selecting a word or phrase (which is usually highlighted) will automatically provide other information about the word or phrase.
Information society	The information society is a term which describes how progress in IT and communications is changing the way people live: how they work and do business, how they study, do research and are trained, how children are educated and how people are entertained.
Interlocutory relief	A temporary remedy, such as an interim payment, granted to a plaintiff by a court pending the trial.
Internet	Technically it is a global "network of networks" connected to each other using common protocols. However it is also a vast source of information in different forms and a way in which people worldwide can communicate with each other.

Internet port number	A number that identifies a particular Internet application.
IPR	Intellectual property rights.
IT	Information Technology
key escrow	The facility for private keys to be held by an independent authority.
NCC	National Computing Centre.
ODETTE	Organisation for Data Exchange by TeleTransmission in Europe. A project undertaken by the automotive industry.
PC	Personal computer.
PRA	Public Records Act 1958.
PRO	Public Records Office.
Privity of contract	The law of contract in general terms only deals with the obligations between parties to a contract and therefore parties who are privy to that contract, or who could be said to have "privity of contract".
S-HTTP	Secure HyperText Transfer Protocol, a secure version of HTTP that uses public key cryptography.
Ultra vires	(Latin: beyond the powers). An act by a public authority, company or other body, which goes beyond the limits of the authority conferred on the body by law and is therefore invalid.
URL	Uniform Resource Locator. A pointer to information on the World Wide Web.
Usenet	A large collection of discussion groups involving millions of people. Each discussion group usually centres around a particular topic.
Virus	A program which relocates itself on computer systems by incorporating itself into other programs which are shared among computer systems.
World Wide Web	Often known as the Web. A hypertext based system for finding and accessing information on the Internet.

Index

To help the reader, this index covers both the *Guideline* and the *Reference Book*. References to the *Guideline* are preceded by the letter 'G' and references to the *Reference Book* are preceded by the letter 'R'. Where an item has references to both books the 'R' references come first.

Printed in the United Kingdom for HMSO
Dd302583 5/96 C10 G3397 10170